D0041662

Frogs, flies,
and dandelions

Frogs, flies, and dandelions

Speciation—The Evolution of New Species

MENNO SCHILTHUIZEN

UNIVERSITY PRESS

OXFORD
UNIVERSITY PRESS

Great Clarendon Street, Oxford OX2 6DP

Oxford University Press is a department of the University of Oxford.
It furthers the University's objective of excellence in research, scholarship,
and education by publishing worldwide in

Oxford New York

Athens Auckland Bangkok Bogotá Buenos Aires Calcutta
Cape Town Chennai Dar es Salaam Delhi Florence Hong Kong Istanbul
Karachi Kuala Lumpur Madrid Melbourne Mexico City Mumbai
Nairobi Paris São Paulo Singapore Taipei Tokyo Toronto Warsaw

with associated companies in Berlin Ibadan

Published in the United States
by Oxford University Press Inc., New York

First published 2001

A catalogue record for this title is available from the British Library

Library of Congress Cataloguing in Publication Data

(Data available)

ISBN 0 - 19 - 850393 - 8
1 3 5 7 9 10 8 6 4 2

Typeset in Bodoni by
J&L Composition Ltd, Filey, North Yorkshire
Printed in Great Britain on acid-free paper by
T. J. International Ltd, Padstow, Cornwall

And after another long time, what with standing half in the shade and half out of it, and what with the slippery-slidy shadows of the trees falling on them, the Giraffe grew blotchy, and the Zebra grew stripy, and the Eland and the Koodoo grew darker, with little wavy grey lines on their backs like bark on a tree-trunk.

Rudyard Kipling (1902), *Just So Stories*

Contents

PROLOGUE

The Making of Species

Archirrhinos haeckelii from the Pacific Hi-Iay islands is a remarkable creature. To the casual observer, it might appear to be a sort of shrew with an exceptionally large nose. When seen in pursuit of prey, however, its uniqueness becomes apparent. Having caught a cockroach, the animal will quickly dive forwards on to its fleshy snout. The snout then spreads out to support the entire body, leaving all legs free for handling and crushing the prey.

This deliberate use of the snout is a distinguishing feature of all Rhinogradentia, the order of mammals to which *Archirrhinos* belongs. Other members, all restricted to Hi-Iay, have developed this specialization even further. *Hopsorrhinus aureus*, the toothed snout leaper, for example, uses its long, flexible nose not only as a support but for actual locomotion. Others, like *Emunctator sorbens*, have modified their snout for feeding purposes. Several individuals of this species will line up along a stream and catch small invertebrates in the mucous discharge from their nostrils.

It might seem odd that these remarkable animals never figure in Sunday-evening wildlife documentaries on television. There are good reasons why they don't. First of all, the Hi-Iay islands are remote, inaccessible, and the Rhinogradentia elusive. Another complicating factor is the complete annihilation of the islands, the Darwin Institute there, and the World Congress of Rhinogradologists by a nuclear disaster in the 1950s. And then, of course, there is the fact that they never existed in the first place.

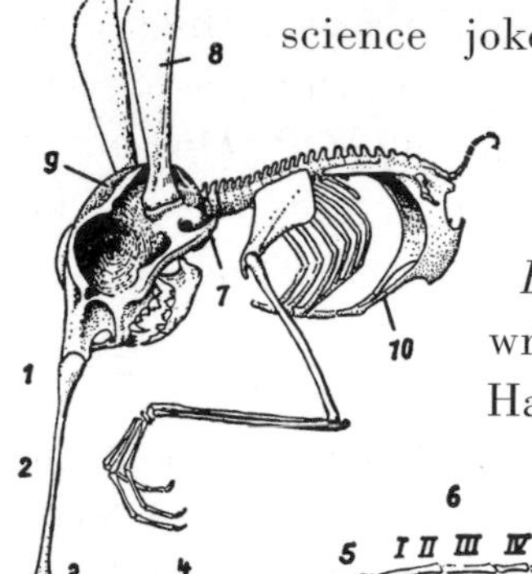

Skeleton of *Otopteryx volitans*. From Stümpke (1957). Reproduced with permission.

For the Rhinogradentia are among the better science jokes of this century. They were invented by Gerolf Steiner, a German zoology professor, who wrote a book about them in 1957. *Bau und Leben der Rhinogradentia*, written under the pseudonym of Harald Stümpke, has been popular among biologists ever since. A certain university lecturer is even known to have used Stümpke's book and kept a supposedly live *Hopsorrhinus* in a cardboard box to play a joke on his students. And every year the more gullible students fall for the trick. To a certain extent this is because Stümpke's work is quite convincing. Even though it includes such obviously fabricated animals as *Rhinochilopus*, which had a nose like a church organ (and to which one tenacious rhinogradologist managed to teach two of Bach's fugues), the book looks exactly like a respectable German monograph. It has all the necessary paraphernalia: anatomical drawings, distribution maps, embryology, a long list of literature, and an epilogue in which the unfortunate loss of the whole archipelago is lamented.

But even more persuasive is the fact that, in all their apparent improbability, there is no reason why the Rhinogradentia shouldn't exist. The world is full of animals that are like them in many respects, differing only in the fact that they really evolved at one time or another, whereas Rhinogradentia have not (yet).

Charles Darwin's famous finches from the Galápagos Islands are a case in point. They did evolve noses (or, rather, their beaks) to suit all walks of avian life. With 13 variations on beak shape, the Galápagos finches have been moulded into an equally large number of different species. One species with a heavy beak specializes in cracking open tough seeds, while more slender-billed versions behave as warblers, catching insects and even drinking nectar. As Darwin noted in 1845, 'One might really fancy that, from an original paucity of birds in this archipelago, one species had been taken and modified for different ends'.

Tenrecs are another, though less famous, example. These mouse-sized mammals, which live on the island of Madagascar, have fanned out into a motley crew of 30 or so species. Some climb, some crawl, some swim, some leap, some burrow. Some produce litters that run into the dozens, some have just one or two young. Some are hairy, some are spiky. Some eat worms, some eat termites, some eat frogs. Some are as large as big rats, some are as small as tiny shrews. But none is identical.

Tenrecs, Darwin's finches, even those Rhinogradentia from Harald Stümpke's imagination, are diverse: biodiverse. Strangely, the word 'biodiversity' itself dates back no further than 1986, when it was invented by some American press officer with a knack for catchy phrases. It is hard to understand how the world managed to do without it for so long, because our environment and our lives are dripping with biodiversity (although, today, a bit less so than they used to).

The most obvious aspect of biodiversity comes in the form of species. 'Species' is just a formalized biological term for what non-biologists usually call 'types', 'forms', or 'kinds'. (A new kind of bird has been discovered . . . Two different forms of palm grow on this island . . . What type of butterfly is that?) And species diversity is staggering: there are 13 species of Darwin's finches in the Galápagos, around 30 tenrec species in Madagascar, 600 bird species in Thailand, 3000 beetle species in Britain, 75 000 mollusc species world-wide, and millions and millions of insect species in tropical rainforests. And that is counting only the animals. Go into your local supermarket and sample the grocery shelves. Kiwi fruit, coconut, cauliflower, carrot: they are all different species. If you didn't smother everything in grass or pebbles, your garden could harbour a hundred different species of plant. And your beer, cheese, wine, yoghurt, and vinegar are all produced by different species of microbe—as are your boils and your diarrhoea.

What are species and why are they there? Why has nature apparently decided to split itself up into different kinds of living things? And why are there so many of them? Does the average European country really need 6000 different beetles and at least twice as many wasps? Why do oak trees in Washington look different from those in Wiesbaden? And why are the mushroom species growing under birches different from those under beeches? Even if there is some perfectly good ecological reason for all this, how did those species come about? How did the finches that got blown all the way from South America to the Galápagos Islands get to be an array of 13 vastly different kinds of birds? How did one reptilian creature give rise to more different dinosaurs than you can turn into plastic museum-souvenirs? And how did a bipedal ape on the African plains split into an array of creatures, one of which eventually came to read popular science books?

'By evolution', is the obvious and correct answer. But what sort of evolution? Imagine a diagram of an evolutionary tree. One of those things that hang in biology classrooms and sit in the inside of a science book cover. If we ignore the worms and jellyfish and rabbits and sharp-dressed businessmen at the extremities of the tree, it is made up of basically two things: the forks where branches split and the branches themselves. But the type of evolution going on in a branch is different from the evolution that produces a fork.

The first—the branch—is the sort of evolution that Darwin wrote about mostly. Natural selection makes animals and plants evolve over time, as they continuously adapt to ever-changing circumstances. A moth in Elizabethan England may have beige wings as a camouflage against the colour of the trees it prefers

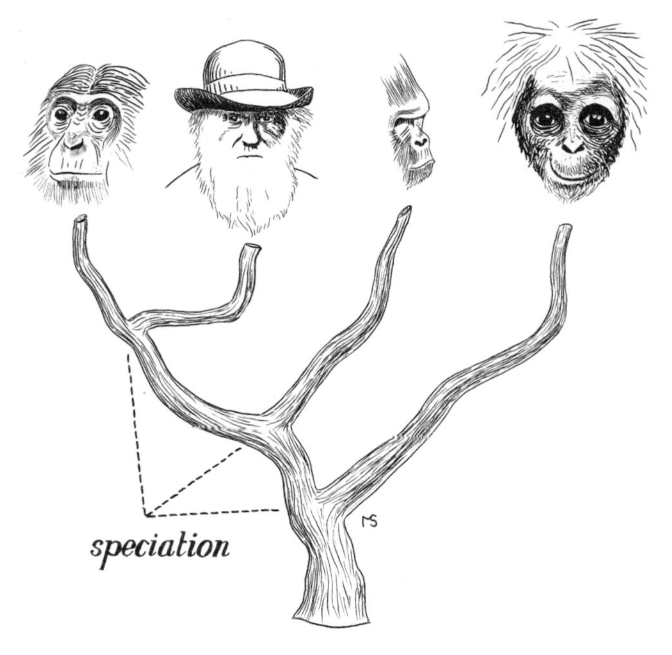

An evolutionary tree of Darwin and the great apes—chimpanzee, gorilla, and orang-utan. Each fork in the tree is a speciation event.

to sit on. After the industrial revolution, as soot darkens everything in the countryside, their beige wings suddenly stand out, making them easy prey for birds. Only the moths with the darkest wings are spared. Over the years, therefore, the species of moth has turned from beige to sooty. Natural selection has adapted it to its altered environment. This is the sort of evolution that a branch represents. It may change, but it remains a single species.

But what happens at the forks is somewhat different. Like in the branches, natural selection may go on in a fork. But rather than changing the appearance of a species, it makes the species split in two. For example, where there first was one species of big cat, there are now two: lions and tigers. Each will go their own evolutionary way. This evolutionary process of splitting, which multiplies the number of species, is called 'speciation'.

Darwin called it by a different name. He spoke of 'divergence of character' and some biologists have said that it was not a subject that he had very clear ideas about. As the famed biologist Edward O. Wilson has complained: '[Darwin's] thinking on diversity remained fuzzy'. And although Darwin as we shall see later in this book, had a better grasp of speciation than most modern biologists think, he still liked to refer to it as 'that Mystery of Mysteries'. Of course, that was in the mid-nineteenth century, when evolutionary biology was in its infancy. But in 1922, the influential biologist William Bateson still said, 'that . . . bit of the theory of evolution which is concerned with the origin and nature of species remains utterly mysterious'. And even as recently as 1974, geneticist Richard Lewontin stated that, 'we know virtually nothing about the genetic changes that occur in species formation'.

All this frustration is understandable. Evolutionary biologists suffer the daily tantalization of not being able to see evolution happen. Of course, they see minor changes, mostly natural selection—like those moths that change colour over a few decades. But that sort of thing (somewhat degradingly called 'microevolution') is peanuts compared with the evolution of entirely new species. Biologists can accept that they will never be able to witness great evolutionary events such as the dawn of the dinosaurs, the origin of humans, or the evolution of birds. But just seeing a single new species evolve, surely, that shouldn't be too much to ask for?

And yet, even though new species must be in all stages of evolution around us, biologists used to think that the entire process must take many thousands of years, and that speciation is therefore much too slow a process for them to study directly. But the past twenty years have proved such pessimists wrong. Speciation is not always so slow. It can be as rapid as a few hundred years; faster even. It can be recreated in a science lab or even in your back garden (under competent supervision, of course). It's big time 'macroevolution'; it's the stuff biodiversity is made of, and it's no longer the Mystery of Mysteries. Nevertheless, although we are beginning to understand speciation events, we owe them respect; they are so endlessly fascinating. Like love stories, they happen over and over again, each time in a subtly different way, depending on the characters of the leading players and on the circumstances, and each is a miracle in its own right. A miracle of miracles.

CHAPTER ONE

Sorting out Life

What are Species Anyway?

If this were a book about chemistry, there would be a good chance that it would start with explaining what chemical elements are. After all, elements are what chemistry is made of. Such a chapter would explain that different elements have different numbers of protons in their atomic nuclei, whereas the number of neutrons might differ. An oxygen atom has eight protons, and any number of neutrons between 6 and 14. It is thc number of protons that determines whether it behaves chemically as oxygen, not the number of neutrons. So an atom with eight protons and eight neutrons is oxygen, whereas an atom with nine protons and eight neutrons is fluorine. Simple. Straightforward.

This is not a book about the periodic table of elements, but about the origin of the biological things called species. 'Species' is a Latin word meaning 'sort' or 'kind'. And species are to biologists what notes are to musicians. Everything in biology revolves around them. Ecologists look at how species interact with their environment. A geneticist compares the genes from one species with those from another. A taxonomist ranks species into families, orders, and classes. Randomly open a book about biochemistry, toxicology, palaeontology, or anatomy, and there is a fair chance that you will find the S-word on the first page you see.

Without species there would be no natural history museums, no field guides to birds, no butterfly collectors, and no fossil hunters. Orchid exhibitions would remain unvisited; *haute cuisine* would become a pointless enterprise; and, on TV, gardening programmes would be more boring than they already are. Conservationists would have nothing left to fight for; Greenpeace could pack up, as could doctors treating infectious diseases. Diversity, *BIO*diversity, the very fabric of life, is built of species.

It goes without saying that biologists know exactly what species are, just like any science teacher will be able to explain in less than a minute how chemical elements are defined. So, as a formality to get out of the way before delving into the juicy bits of evolution, here goes: species are animals or plants that look like each other. They belong to different species if they do not look like each other. Lions and tigers are different species, because the latter is stripy, the former is not. The dandelions in your front garden and the ones in your neighbour's garden, on the other hand, obviously belong to the same species, because they look the same. There you have it: we have defined what species are. Now we can embark on our tortuous journey into tropical jungles, colourful birds, the wonders of genetics, and the treasures of the world's greatest natural history collections.

Not so.

If only it were true . . .

As a matter of fact, there are few questions in science that have elicited such a long-standing and heated debate as the species problem (sometimes spelt with capital initials to indicate the size and importance of the issue, hence The Species Problem). Dozens of books, hundreds of doctoral theses, and thousands of scholarly articles have been devoted to the subject. A plethora of species concepts and definitions have been proposed over the years and the number of disputes has grown exponentially with them. Tears have been shed, fists have been clenched, and careers have been broken, all as a result of the simple question, 'What is a species?'.

Sometimes, this question is augmented with the sarcastic clause 'if anything'. 'What, if anything, is a species?' Usually this is asked by philosophers who suspect that biologists and naturalists are just deluding themselves into thinking that the species concept is real. They prefer to think that there is just gradual variation, and that biologists arbitrarily draw lines between species, just to make the hotchpotch of biodiversity manageable. But anybody who has ever played the game of counting waterfowl species in the local park pond knows this cannot be true. Even birds that look very similar, such as mallards, pintails, shovellers, and gadwalls, can be easily told apart and, what's more, different people feeding the same ducks will agree on the different categories.

Perhaps even more convincing is the fact that tribal people who have never seen a naturalists' field guide usually distinguish the animals and plants around them in the same way that a taxonomist would. For example, in 1928, ornithologist Ernst Mayr went on an expedition to the Arfak mountains of New Guinea and asked native hunters to bring him bird specimens. For each specimen that was delivered to him, he asked his helpers what it was called in their language. In the end, Mayr had to conclude that his job had already been done for him: the Arfak people recognized exactly the same 136 bird species as he, the experienced Western taxonomist.

Keep it in the family

But even if we agree that species are real things, how do we decide which organisms belong to the same species and which belong to different species? Most of the time, the look-alike criterion seems to work fine. A llama is the spitting image of another llama; two peas in a pod are as alike as they can be; if you've seen one starling, you've seen them all; and a horse is a horse is a horse, of course. Of course? Then what about Arabian stallions, Shetland ponies, and the 34-inch miniature horse? They certainly do not look alike, and a taxonomist from Mars would definitely classify them as different species. And he would do the same with our hundreds of breeds of dogs (ranging from chihuahuas to great danes) and with President Kennedy and Marilyn (two tulip varieties).

Our Martian taxonomist would encounter problems in the wild, too. One common European snail, the grove snail, for example, has a shell that comes in three colours—brown, pink, or yellow—and it can be decorated with up to five black stripes. In grassland, yellow stripy ones are well camouflaged, so a hungry thrush will eat only the brown and pink ones there. In shadowy woodland, though, the situation is reversed: here, the yellow-and-black form stands out, so birds will more likely make a meal of those. As a result, the Martian visitor on a field trip will distinguish three species of grove snail: a yellow one with black bands that lives in grassland, and a brownish and a pinkish one that live in woods. On the same visit, he might encounter more problematic specimens: a male stag beetle with its antler-like jaws looks nothing like its female, so he would probably view them as two species. And, again, based on appearance alone, our Martian would never lump a moth with its juvenile stage, the caterpillar.

Nevertheless, any Earth taxonomist could point out the mistakes to his extraterrestrial colleague. Artificially selected breeds of horses all belong to a single species, *Equus equus*, and all dogs—great, small, long-haired, short-haired, pure-bred, or mongrel—are *Canis familiaris*. Along similar lines, natural selection also produces 'breeds', by making the grove snail *Cepaea nemoralis* take different forms in different surroundings. Male and female stag beetles are both classified as *Lucanus cervus* and the larval and adult moth likewise belong to one and the same species. Why would this be? Apparently we do not just use the look-alike feature for grouping organisms into species. It works often, but not always. So what other criteria can we use? The answer is not difficult: it's breeding. Male and female stag beetles mate with each other and produce sons with gigantic jaws and daughters with small jaws. Likewise, any pair of dogs, however different they may look, can (at least in theory) mate and have puppies.

The problem with the look-alike criterion. The donkey (left) and horse are two different species, whereas all races of domestic dog belong to the same species.

So interbreeding and looking similar are two sides of the same coin. Animals and plants that normally mate with one another share, loosely speaking, the same gene pool. As a result, they will all more or less look like each other. If they don't, it means that they are not the same age or sex, or that natural or artificial selection has singled out a particular appearance for particular circumstances.

Many biologists have used this flip side of the species coin to make formal definitions of what species are. As early as 1849, the German evolutionist Albert Mousson wrote: 'the species is the total of individuals, interconnected by descent and reproduction, maintaining unlimited reproductive capabilities'. The most famous definition has come from Ernst Mayr, whom we've already met, trawling through deceased New Guinean birds. Mayr is a great biologist who dominated evolutionary biology for most of the twentieth century and whose name and ideas will come up regularly in this book. In 1940, Mayr wrote that species are 'groups of actually or potentially interbreeding natural populations which are reproductively isolated from other such groups'. He termed his definition the 'biological species concept', abbreviated to 'the BSC'.

Mayr's biological species concept seems like a nice, simple definition for what species are. No matter how different two organisms may look, if they can interbreed, they share a common gene pool, and, by definition, they belong to one and the same species. But the apparent clarity and practicality of the BSC are deceptive, because nature is not as simple as Mayr's definition. Jim Mallet, an evolutionary biologist of University College London, wrote in 1997: 'Mayr's heritage is a huge muddle. By postulating an ideal species . . . rather than a practical approach to sorting actual taxa, Mayr opened a Pandora's box.'

Into the mongrel zone

One of the problems is that gene pools are not always watertight: seepage, leakage, and downright oozing of genes occurs. Around 12% of European butterfly species hybridize among each other, and about 25% of all American warblers do so. Even the blue whale, the largest animal that ever lived, is in the club, and occasionally takes to breeding with the slightly smaller fin whale. And everybody knows that useful love-child of horses and donkeys, the mule. But mules, and other products of adventurous liaisons across the species border, are sterile and cannot breed. So, in spite of bastardy, donkeys and horses more or less comply with the BSC, remaining in separate gene pools.

The same cannot be said, however, for fire-bellied toads. In Europe, two species of these amphibians occur; and Polish herpetologist Jacek Szymura from the Jagiellonian University in Cracow has built his 20-year career on them. Szymura, a wiry man with Groucho Marx mannerisms and moustache, who draws crowds of amused students for his amphibian impersonations at conferences, knows all there is to know about fire-bellied toads and more. In Western Europe, he says, lives *Bombina variegata*, which is yellow-bellied and frequents small ponds in mountainous areas. The males are not territorial and have a very soft, rapidly repeated call ('Oob-oob-oob-oob-oob', Szymura goes). The Eastern European *Bombina bombina*, on the other hand, has red blotches on its belly, breeds in large permanent waters in the lowlands, and has aggressive, territorial males with a loud mating call, amplified by an air sac ('Whooomp!, Whooomp!'). Carolus Linnaeus, the eighteenth-century Swedish naturalist, noticed the difference in their call, remarking that the 'German frogs croaked very loudly . . . so mournfully that one almost died of melancholy'. And these are just a few of the 20 or more distinguishing marks that Szymura has on his list. 'I would use the term species for them', he says.

Nevertheless, the fire-bellied toads stubbornly refuse to comply with Mayr's no-interbreeding rule. Along a line less than 10 kilometres wide and 5000 kilometres long, running from eastern Germany all the way down to the Black Sea, Szymura discovered that the two species hybridize. And they do not just hybridize the way horses and donkeys do, forming the occasional sterile bastard. No, the species mate, form hybrids, which in turn mate among themselves and with the parent species, the offspring of which also join the pool of sexual interchange, and so on and so forth. As a result, the 'hybrid zone' is a genetic no man's land filled with all kinds of intermediate toads, mixtures of the two 'species' on either side.

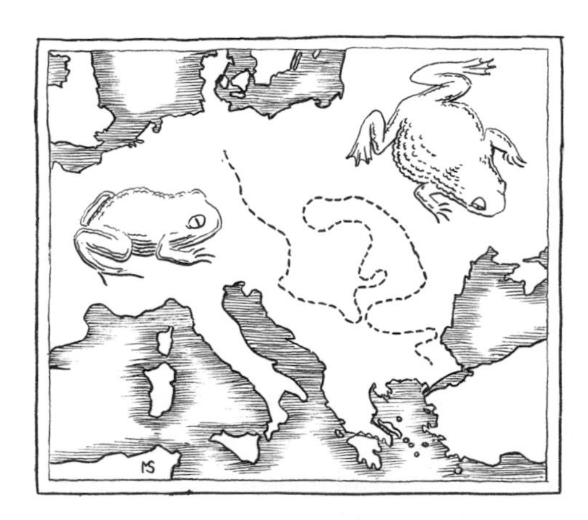

The two fire-bellied toads, *Bombina bombina* and *B. variegata*, and the hybrid zone in central Europe, where they interbreed.

Such hybrid zones are not rare. Hundreds of examples exist. Depending on the animal's or plant's abilities for movement they may be narrow or wide. At least 10 different pairs of bird species form hybrid zones up to 1000 kilometres wide in the Great Plains of North America, while proverbially slow-moving land snails have hybrid zones as narrow as 50 metres across on the island of Crete.

Taxonomists are a clever lot. Faced with the conflict between what their gut feeling tells them (animals or plants that look different, live in different biotopes, and have differences in their behaviour must belong to separate species) and the uncompromising rule of the BSC (if it interbreeds, it's the same species), they have pulled the ultimate taxonomist's trick: subspecies. An odd taxonomic category, subspecies are subdivisions of species, based on essentially the same criteria that taxonomists use to separate species. They will look different, live in different parts of the world, and often have differences in their habits and habitats. Unlike 'real' species, however, they do not pass the acid test devised by Ernst Mayr: they can interbreed. Subspecies even have formal scientific names, made up of the well-known double name, with a third name tagged onto it. So, strictly speaking, the fire-bellied toads should be called by the somewhat pleonastic names of *Bombina bombina variegata* and *Bombina bombina bombina*.

In fact, many of the pairs that form hybrid zones were initially classified as distinct species, until the hybrid zone was discovered and they were duly degraded to subspecies rank. For example, the black-and-grey, eastern European hooded crow and the all-black, western European carrion crow were known until the early twentieth century as *Corvus corone* and *Corvus cornix*, respectively. After having mapped the hybrid zone that runs from Scotland and Denmark through the Alps down to the Mediterranean, ornithologists grudgingly assigned them subspecific status and they now go by the names of *Corvus corone corone* and *Corvus corone cornix*.

BSC without honours

Yet the biologists' habit of using subspecific names for distinct organisms that hybridize exposes the Achilles' heel of the biological species concept—namely that it is not biological at all; instead it is philosophical. Mayr and his followers defined an 'ideal' species and then proceeded to try to fit nature into that ideal. It is true that many of the species that taxonomists distinguish mate only among themselves, but the BSC turns this upside down and claims that species 'should' be genetically isolated to qualify. And the closer biologists look, the less comfortable they feel with the BSC.

Jacek Szymura, for example, recorded the individual belly patterns of his fire-bellied toads (whose bellies differ from toad to toad as much as human faces do from person to person). Having ID-ed the animals this way, he then could track each one's movements over a year. He discovered that they do not stay put. The toads travel so much that, on average, they lay their eggs about half a kilometre from where they themselves were born. With this capacity for movement, and hence for meeting members of the opposite species, the two forms should have merged like two splashes of water colour.

Yet, the zone is only a few kilometres wide and it does not seem to be getting any wider. How can this be? Apparently, in spite of all the hybridization that goes on, the *bombina* and *variegata* species manage to keep their identities. Szymura has evidence that the toads whose genetic make-up is a mosaic of genes from both species—the mongrels that inhabit the hybrid zone—are less healthy than the 'pure' individuals. The poor bastards often die as embryos, and even the ones that survive to adulthood frequently suffer from deformities (ribs fused to vertebrae, lopsided bodies, or abnormal mouths in tadpoles). So even though there is a lot of hybridization, natural selection steadily filters out many of the mongrels, which prevents the area where hybrids form from spreading out. So the two species continue to live literally side by side.

Even if scientists cannot directly prove that hybrids suffer from poor health, they sometimes have circumstantial evidence that two species can hybridize for thousands of years without amalgamating into a single gene pool. Satoshi Chiba, for example, a zoologist from Shizuoka University in Japan, has been doing field work since the late 1980s on the Bonin Islands (Ogasawara-shoto), a group of tiny volcanic islands in the Pacific, more than 1000 kilometres off the coast of Japan. On these 'Japanese Galápagos islands', Chiba has been studying escargot-sized snails called *Mandarina*. On one island in the archipelago, Hahajima, two species occur, *Mandarina ponderosa* and *M. aureola*. To be absolutely sure which is which, Chiba always examines the snails' sex organs, but colour also is quite a reliable way of identifying them: *M. ponderosa* is dark brown, with two black bands running round the shell, while *M. aureola* is bright red or yellow, with three black bands. Where the two species live together, hybrids occur that are intermediate: lighter brown with two bands and a faint third one, darker brown with two bands, or some other combination. When Chiba checked the genes of snails from such places, it was clear that the hybrids are not sterile: genes from one species flow, via the hybrids, into the other species.

This hybridization between the two species is not just an opportunist fling: they have been at it for thousands of years. In the dunes along the coast of Hahajima, Chiba discovered layers of old shells, embedded in clay and sand. Using the well-known radioactive carbon method, he dated the fossil shells to roughly 2000 years old. But there was one snag: the bleached shells had lost all their colour and, of course, their sex organs as well, so how could he decide to which species they belonged? Palaeontologists have a neat trick for this. If ultraviolet light is shone on colourless fossil shells, they momentarily regain their erstwhile colour patterns. So Chiba did this and saw the double bands of *M. ponderosa*, the triple bands of *M. aureola*, and, indeed, the intermediate colours of hybrid shells. Once again, two species had managed to keep their own identities in spite of thousands of years of interbreeding.

Hybridizing species are not the only felons that break the laws of the BSC. There are more problems. What to do with so-called species that live in different regions and never meet each other? The forests of Peninsular Malaysia, for example, are spotted by some 300 limestone hills, rocky outcrops each the size of a block of flats, which rise precipitously from the surrounding jungle. The transformed and uplifted remains of ancient coral reefs, they stand isolated, sometimes hundreds of kilometres away from the next hill. And yet, long ago, tiny limestone-inhabiting land snails managed to seek them out and now flourish there. But because these hill-colonies have been isolated from each other for millions of years, many of these snails have changed in shape and have slowly evolved into hundreds of different species. Or that is how malacologists (mollusc-studying biologists) consider them.

But, of course, these malacologists do not check if snails from different hills fail to mate with each other. They just notice that the snails on one hill have spikes on their shells, whereas the ones on the next hill have knobs, and the snails on a third hill have neither. Hence, the snails belong to different species. Malacologists studying the Malaysian limestone hill fauna simply apply the look-alike criterion, and for good reasons. After all, with hundreds of different candidate species, one would need to do tens of thousands of crossing experiments to determine whether they belong to different 'biological' species or not. And even if this were possible, the experiments would be pretty meaningless, because the snails never meet each other in nature, so they are already 'genetically isolated'.

So, in all fairness, Mayr's biological species concept is not very useful. Things that every biologist would happily classify as distinct species can engage in unbounded hybridization in hybrid zones. And with animals or plants that do not live in the same place, there is no way of telling whether they comply with the

BSC or not. Nevertheless, the BSC dominated biology for most of the twentieth century and it has appeared as the standard definition of a species in many textbooks. It has even been adopted as the official species definition in conservation legislations, such as the Endangered Species Act in the United States.

Back to Darwin

If the look-alike concept and the BSC do not work, what alternatives are there for defining a species? There are about a dozen or so. Over the years people have emphasized all kinds of aspects of what it means to be a species, and elevated them to definitions. Some have said that species must be descendants of a single common ancestor. Others, in desperation, have decided to let the organisms themselves sort it out: if they recognize each other as sex partners, they belong to the same species.

In evolutionary hang-ups, it is always worthwhile to go back to the old master for sound advice. What did Darwin have to say about this? After all, Darwin thought and wrote a lot about species and how to define them. *On the Origin of Species* was the culmination of 30 years of philosophizing about species, what they are, and how they evolve (in fact, it was the rushed overture to a much larger 'Species Book' he had in mind but never finished). He even wrote a taxonomical monograph on barnacles, as an eight-year act of self-discipline to learn the difficulties of defining and delineating species.

Darwin struggled long and hard with his subject and over the years more than once changed his thinking. In the 1830s, when he had just returned from his round-trip of the world, he apparently adhered to one of the early versions of the BSC. He noticed that, due to a certain 'repugnance', individuals of different species would not mate with each other, writing that 'until their instinctive impulse to keep separate' is overcome, 'these animals are distinct species'. But by the time he wrote *On the Origin of Species*, 20 years later, he had changed his mind and now felt that the term species was 'arbitrarily given for the sake of convenience to a set of individuals closely resembling each other'. In a private letter he called the entire business of trying to find the essence of species a 'laughable' attempt at 'defining the undefinable,' and in the *Origin* he simply said: 'the opinion of naturalists having sound judgement and wide experience seems the only guide to follow', adding that 'this may not be a cheering prospect, but we shall at least be freed from the vain search for the undiscovered and undiscoverable essence of the term species'.

In philosophers' terms, Darwin had given up on essentialism and opted for pragmatism instead. He had decided, after 30 years of studying animals and plants and testing in his mind various definitions and concepts, that there is no 'essence' that sets species apart from the rest of natural classifications. Rather, his taxonomic training told him, just like anyone who goes for a walk in the park and who keeps a mental list of the different birds that can be seen in the pond, that species are just groups of animals or plants that share a bunch of characteristics with each other. No more, no less. This may smack of the cynicism of the armchair biologist, who denies that there are species in the first place. But there is a difference. The conclusion Darwin reached was that species do exist and the gaps between them are real, but that it is impossible to set up one single golden standard that tells what makes a species a species. To Darwin, they were no more than 'well-marked varieties'.

As Darwin anticipated, this stance at first seems a bit disappointing. Somehow scientists and park-visitors alike have the intuition that species must have an undiscovered core essence that makes them something special, and we feel let down by Darwin. He, more than anybody else, should have realized that. Ernst Mayr even scorned Darwin for his cowardly turnabout and wrote in 1963: 'Darwin's failure . . . resulted to a large extent from a misunderstanding of the true nature of species.' But many biologists now think that it was actually Mayr who was confused, when he felt the need to capture the 'true nature of species' in such an unworkable definition as the BSC. In fact, science is at present experiencing a modern version of the change of heart that Darwin had between the 1830s and the 1850s. It may seem ridiculous that it has taken biology one-and-a-half centuries to come full circle, but there is a good reason why this is happening now, and not sooner. Biologists today have the tools that Darwin did not have. And those are the tools of genetics.

The age of genetics

While experimenting with plant hybrids, the Silesian monk Gregor Mendel discovered how heredity worked in 1865, only six years after *On the Origin of Species* was published. But his research was ignored and heredity remained a mysterious business during the entire nineteenth century. Mendel's paper was rediscovered only in 1900, and, up to that time, theorizing about genetics had become something of a cottage industry. Darwin himself invented the theory of 'pangenesis'. This involved so-called gemmules, which circulate in the body, picking up information about what their body looks like and then gathering in the reproductive organs where they could be transmitted in the form of eggs and sperm. One of the consequences of pangenesis was the inheritance of what are called 'acquired characteristics'. Darwin, and in fact most of Victorian society, wrongly believed that, for example, 'the domesticated rabbit becomes tame from close confinement; the dog intelligent from associating with man; . . . *and these mental endowments . . . are all inherited*' (italics added).

We now know that this is not true. A molecule called DNA is responsible for heredity. It carries thousands of genes, which together form the blueprint for how the organism looks and functions. DNA is not changed by the rest of the body, which is why tameness or other things learnt during life cannot be passed on to offspring.

Normally a plant or animal carries two copies of each gene. One is inherited from its father, the other one from its mother. If these copies are different, these versions of the same gene are termed 'alleles'. And even though genes are linked together in long chains called chromosomes, these connections are broken up into smaller blocks when the sperm and eggs are formed, during a process called recombination. As a result, every individual carries a unique and random combination of its parents' genes.

To visualize recombination, think of an animal with only two chromosomes: one from its mother and one from its father. Now imagine these chromosomes as long trains with lots and lots of carriages: the genes. When this animal produces a sperm or egg cell, it aligns its two gene-trains on parallel tracks. Railway workers open the connections between the carriages (the genes) at a number of places along the entire length of both trains.

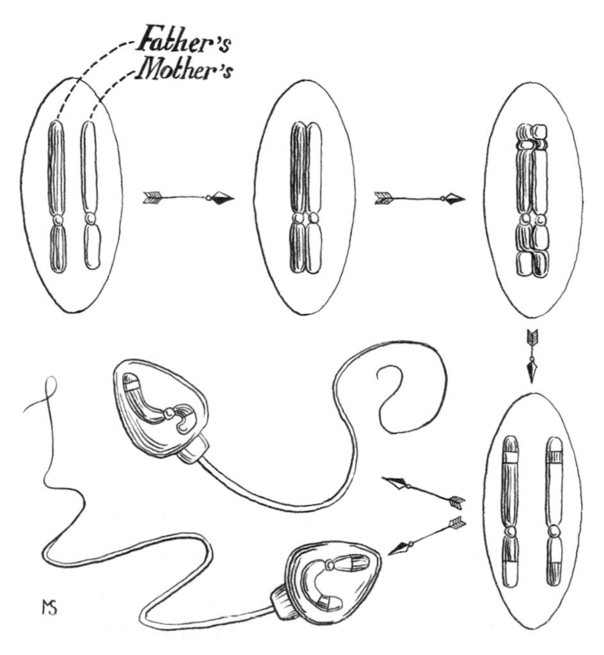

Simplified diagram of recombination. Before chromosomes are packed singly into sperm (or egg) cells, they exchange parts of their arms.

Next, they interchange the loose stretches of carriages between the two trains. Then each newly assembled train rides off to the next generation as a sperm or egg cell. In the next generation, the same happens. But because the places where the railway workers disconnect carriages are a bit haphazard, the gene-trains keep changing composition. In the end, after many generations, no carriage (gene) sits in between the two carriages that were its neighbours in its great-great-great-grandparent. So to all intents and purposes, genes are inherited separately.

If we broaden our field of view to encompass a lot of interbreeding individuals, it becomes clear why biologists like to speak of the 'gene pool.' Because of all this sexual mixing, genes behave in a population in much the same way as molecules in liquids do. It may seem far-fetched, but a history of kitchen and bathroom doodling can give one an intuitive understanding of population genetics. For example, migration is often called 'gene flow'. Imagine that two populations of a particular plant grow on either side of a mountain range. Let us assume that one population has red flowers, the other white flowers. Every flowering season, the prevailing winds are such that clouds of pollen grains with alleles for red flowers waft over the mountain tops and land on the stamens of the white flowers on the other side. In time, this 'gene flow' will eventually redden the white population, just as seepage can turn fresh water behind coastal dykes brackish.

In a similar vein, we can try to visualize the action of natural selection. Suppose a plant with the allele for white flowers is actually superior to one with the allele for red flowers (for instance, because the eyes of pollinating insects pick out white much better than they do red). A plant with white flowers will then have more offspring than a plant with red flowers, and, as a result, only the tiniest trickle of white alleles flowing into the red gene pool will kick-start the evolutionary flip to an all-white population. Again, biologists often refer to natural selection in terms borrowed from hydraulic engineering. But because natural selection is more powerful than just the random diffusion of gene flow, they use correspondingly violent expressions, such as 'selection pressure' or 'waves of advance'.

A fascinating example of such a wave of advance comes from research by French geneticist Michel Raymond from the University of Montpellier. During the 1980s, pest-control workers all over the world noticed that mosquitoes of the species *Culex pipiens* started to acquire resistance against organophosphates, which are insecticides like parathion. Raymond pinpointed the gene responsible for the resistance and found that it produces an enzyme that helps break down the pesticide. He then looked at the exact DNA structure of the gene and discovered that all mosquitoes, were they from California, Sri Lanka, or Japan, carried the same allele of the gene. Because it was impossible that the successive mutations that had produced this allele would have occurred more than once in exactly the same way, it could mean only one thing: a mutation had produced the resistance-conferring allele somewhere and selection pressure had pushed it on a wave of advance across the globe.

This had happened with blatant disregard for species barriers, by the way. The mosquito populations of France and Italy, for example, belong to different 'crossing types'. These mosquitoes cannot interbreed, and, in compliance with Mayr's BSC, they should be considered separate species. But this does not seem to have stopped the wave of advance. All that was needed was a single successful fertilization among the millions and millions of mosquitoes of each type for the gene to cross the species barrier and to continue its rampant genetic take-over. So here we have another reason why species concepts can never be based on genetic isolation or on ancestry alone. Genes that evolve in one species, but which related species could also put to good use, will eventually pass into them. Even the tiniest amount of hybridization is enough to let the genie out of the bottle and make this happen.

Leaky genes

Even if there is no hybridization, however, it is now known that the DNA of many animals contains bits that can sometimes run wild and cross species borders of their own accord. These so-called endogeneous retroviruses can jump free from the chromosomes by making a loose copy of themselves, which will self-assemble into a virus. The virus can then infect another animal (not necessarily of the same species), and insert and settle itself into its chromosomes, after which it might again lie dormant for a long time, happily inherited in the conventional way.

This may sound like science fiction, but it happens. For example, all Old World monkeys (that is, the monkeys of Europe, Africa, and Asia) have in their chromosomes a block of DNA that is known as a 'type-C virogene'. This sequence of DNA occurs nowhere else in the animal kingdom, except in the chromosomes of a few species of cat, such as the African wildcat, the jungle cat, and, yes, it also lurks in your domesticated kitty on the couch. Some 5 to 10 million years ago, in the monkey that was the ancestor of modern baboons, the gene must have turned itself into a virus and somehow infected an ancestral cat. Once in the cat, the gene went dormant again and remained so ever since, even while its host split into the six cat species that now carry this tiny bit of monkey DNA.

This shows that chunks of DNA can shuttle from one species to another, so even if they never engage in any hybridization, it is not certain that two species are completely genetically isolated. So one may wonder whether Mayr's ideal of unconditional isolation ever really exists in nature. Absolute genetic barriers are as rare as a cheap tent without leaks.

Some species even appear to make genetic leakage their trademark, such as two of Africa's most dangerous animals—not lions or rhinos but the malaria-transmitting *Anopheles gambiae* and *A. arabiensis*. Medical entomologists scrutinizing mosquito behaviour have uncovered a whole range of subtle but often fatal differences between these two species. Although the two mosquitoes bear an uncanny resemblance to each other, *A. gambiae* is the more deadly of the two. It can carry up to 15 times more malaria parasites in its body, and is distinctly partial to biting humans rather than animals. One entomologist once recorded a swarm of these mosquitoes passing through a herd of grazing cattle only to head straight for the huts of the herdsmen. In villages, *A. gambiae* is also much more likely to look for victims inside houses, whereas *A. arabiensis*, if it attacks people, will do so outside.

Geneticists suspect that the genes responsible for all these differences are crammed together on a little piece of chromosome called the 'Xag inversion'. The orientation of this runt of genes is reversed in *gambiae* when compared with *arabiensis*. The result is that if the two species hybridize, all their chromosomes can align with their partner from the other species, but this Xag inversion cannot. So in the hybrid's offspring, genes from *gambiae* and *arabiensis* will be jumbled, except for those crucial species-determining genes. They will always stay together. The upshot is that, via hybridization, vast parts of the mosquitoes' chromosomes are finding their way into the other species, with the exception of this vital gene cluster. In other words, these species are willing to exchange all the genes they have, as long as they can keep just a small packet of genes for their own private purposes.

Ironically, Mayr was one of the first people to develop and promote this view of, what he called, 'co-adapted gene complexes'. But he thought these frameworks were very fragile and their integrity would be compromised immediately if they were to engage in gene flow with other species. That is why he thought that they needed the 'protection' of barriers against interbreeding. But it now seems that, in malaria mosquitoes and many other species as well, many genes can be replaced by counterparts from a related species without any negative effects. This is why many species are not easily 'undone' by gene flow.

Evolutionary biologists are becoming attracted by such a gene-cluster species concept. Not only does it better reflect what actually goes on at the genetic level, but it also helps to clear up the logjam about the BSC. Some species pairs simply differ by only a few crucial genes, and many genes can be shuttled to and fro by hybridization without disrupting these gene clusters. Other species pairs differ by many genes and there is hardly any region on the chromosomes where no crucial genes lie. Such species pairs are quite resistant to gene flow, and it is these species that comply with the BSC. So in the end, in a genetic sense, Darwin was right: gene clusters may be anything from complex to simple, and they may be more or less resistant to gene flow, but there is no 'essence' of a species: species are just particularly stable combinations of genes.

Mayr (aged 95 at the time of writing) still stands by his BSC. Time for a new species concept? 'No, no, no-no-no', he says over the phone, in English with a German accent that sounds decidedly unyielding. 'A species is a reproductive community!' But even Mayr has not been able to tackle this final problem. What about organisms that don't do sex?

When the gene pool freezes over

Remember the dandelions from the beginning of this chapter, that look the same everywhere and thus, by the look-alike principle, are all one and the same species? Everybody knows dandelions: they occur all over the world as cheerful, almost vulgar yellow-flowered, big-leaved, hollow-stemmed weeds in meadows and along roadsides. They may be a nuisance to gardeners, but who would expect them to be a veritable nightmare to botanists? For dandelions are all alike, yet all different, and they are not one species.

Carolus Linnaeus, who, in the 1750s, invented the naming system for species that we still use today, was oblivious of any pitfalls when he lumped all dandelions together in the species *Leontodon taraxacum*—literally the lion's tooth (the English word 'dandelion' is derived from the French *dent de lion*, which means just that). The problems started when successive generations of botanists began to say that Linnaeus had overlooked small, but consistent differences in the shapes of the leaves and seeds of dandelion. In 1838, a Swiss scientist thought he could tell 30 different species apart; an Austrian, 70 years later, topped that by increasing the number to 57.

But dandelion-taxonomy only really got out of hand in the twentieth century. A band of mainly Scandinavian botanists started the splitting in earnest. Using detailed analysis of minute differences in the leaves, the flowers, and the seeds of the plants, they realized that the previous attempts at distinguishing species had been gross underestimates. Whole stacks of scientific papers were produced, filled with descriptions and names of new species, resulting in a botanical cottage industry that in the end churned out a grand total of no less than 2000 species of dandelions world-wide.

Less-ambitious botanists were bewildered, and most of them still are, faced with the sheer impossibility of simply identifying an ordinary dandelion. Peter van Dijk, an evolutionary botanist at the Institute for Terrestrial Ecology in Heteren, the Netherlands, says:

> It is a hopeless task. In Europe, there are only three or four 'gurus' who say they can identify dandelions, and when you go into the field with them it turns out that even they are already happy when they can get tags on ten percent of the plants they encounter. Even worse, if you send the same plant to two different gurus you get two different answers, and when you send the same plant to the same guru twice, you also get two different answers. Dandelion taxonomy is a very dubious enterprise, I think.

So what is it with dandelions? Why do they stubbornly defy the efforts of well-meaning botanists? The answer lies in their chromosomes. For these plants have evolved a peculiar way of reproducing. Some primitive species of dandelion behave like any other regular sexual plant. They have two sets of chromosomes, which, in their yellow flowers, they split into single sets to produce ovules and pollen. With these they fertilize each other. But most dandelion species have abolished sex altogether. They have three sets of chromosomes rather than two, which means they cannot make normal pollen and eggs, because a triple set cannot be evenly split down the middle. Nevertheless, they do make flowers with ovules and pollen, but the pollen is sterile, and the ovules do not attempt to split their chromosomes. Instead, the ovule cells hang on to their triple chromosome sets and develop into seeds without any intervention of pollen. So they practise parthenogenesis, the biological term for virgin birth. The upshot is that these plants clone themselves. Genetically, the parachute-seeds of a parthenogen are identical copies of the mother plant. They are all Dolly-dandelions.

Sitting side-by-side on a bench in the sunny courtyard of their institute, Peter van Dijk and his graduate student Peter van Baarlen can view the dozens of flowering dandelions that are scattered over the badly tended garden. 'In the Netherlands and other northern European countries, most dandelions are of the parthenogenetic kind', says van Baarlen. 'I have checked some of the dandelions growing here between the pebbles, and they all have irregular pollen, which is a sure sign of parthenogenesis.' So are they all clones of each other? 'Probably not', van Dijk butts in. 'We have been doing DNA-fingerprinting—a technique similar to that employed by forensic scientists—on five hundred plants from a single pasture, and we found sixty different fingerprints.'

So, apparently, clones have evolved many times independently from sexual ancestors. This would explain why botanists think they see so many species. A parthenogenetic plant and all its clonal offspring are a frozen combination of genes, rather than a gene pool that is continually mixed by gene flow. A botanist faced with thousands of dandelions representing only a few dozen of such gene combinations would immediately notice that this was no ordinary collection of genetically varying plants. It is like looking at the group photo of an identical twins convention.

So are botanists right or wrong in distinguishing so many different species? In one sense, they are right, because each parthenogenetic dandelion 'strain' represents a single, probably quite successful, cluster of genes. When we view species as stable gene clusters, each parthenogen is a species. But, on the other hand, each strain is no more than a multiplied individual out of the sexual species that it harks back to. So in that sense, all parthenogens should be lumped with the sexual species from which they are derived. In dandelions, van Dijk thinks that any concept of species breaks down. 'You get a fuzzy sort of evolution with parthenogens splitting off all the time', he says. In fact, he and van Baarlen have discovered that parthenogens on rare occasions can produce fertile pollen, with single sets of chromosomes. They suspect that sexual plants sometimes get fertilized with this pollen, so there is a tiny trickle of gene flow from parthenogens back into the sexual gene pool, which confuses matters even further.

The whole dandelion contention reflects the taxonomic quagmire of other organisms in which cloning is rife, such as brambles, roses, and hawkweeds, over which similar conflicts have started. This goes to show that species as we know them in 'regular' plants and animals exist so clearly only because sex mixes and blends their gene pools. As soon as sex is abolished, the homogenized gene-clusters shatter into myriads of better or worse 'frozen' gene combinations. Without sex, biodiversity would not be partitioned into species and the world would look entirely different.

Now that we have an angle on what species are, why we can recognize them, and why we sometimes cannot recognize them, it is time to start wondering how new species evolve. Why do gene pools split in two and begin evolving in different directions, starting a process that we call speciation? One way is by piggybacking on the forces of geology.

CHAPTER TWO

An Isolated Case?

Geographical Speciation

When Charles Darwin first set foot on the Galápagos Islands, in September 1835, they must have seemed like paradise to him. He had spent the austral winter braving snowstorms and earthquakes in the barren mountains of the Andes, so the 26-year-old naturalist must have welcomed the tropical flora and fauna that the oceanic archipelago offered to him. In the five weeks that he and the crew of HMS *Beagle* spent there, Darwin took to unbounded collecting and experimentation. He was fascinated by the extreme tameness of the birds. In his report he wrote: 'A gun is here almost superfluous; for with the muzzle, I pushed a hawk off the branch of a tree.' The sea iguana appalled him on account of its being 'hideous-looking', 'stupid', and 'sluggish', and he even spent some time repeatedly hurling a particularly unlucky individual into the waves to see if it would eventually flee to sea (it never did, but doggedly kept returning to the shore).

Darwin was also most amused with the animals that give the archipelago its name, the gigantic *galápagos* or tortoises that roamed the islands, and which Darwin rode, ate, and—when he was very thirsty—drank the urine of ('only a *very slightly* bitter taste'). When the vice-governor of the Galápagos, a Mr Lawson, told him that the tortoises from each island were different, and that he could readily tell by their looks from which island they were brought, Darwin would not believe him at first. 'I never dreamed that islands, about fifty or sixty miles apart, and most of them in sight of each other, . . . would have been differently tenanted', he wrote. In fact, Darwin had already lumped the specimens he had collected, without labelling them as to the exact island from which they came. Later he was able to confirm Lawson's claim. Some islands harboured smaller tortoises, some larger ones (on a particular island they grew so large that it took six to eight men to lift a single male). Certain islands had smooth-shelled tortoises, whereas on others the shells had ridges and furrows. The ones from Pinta Island had a shell with its front edge upturned.

Being forewarned, Darwin was more careful with the herbarium he put together. He systematically noted the native island for each plant that he collected and dried. Ten years later, when his botanist friend Joseph Dalton Hooker had worked up his collection, he was able to see that what was true for tortoises was true for plants on a much more massive scale. Of the 71 different plant species that he brought home from Isabela Island, for example, 30 were exclusive to Isabela, and not found on any of the other islands, which had their own aboriginal species. On average, 40% of the flora of each island was unique to that place. The same phenomenon was visible among the birds of the Galápagos—the well-known Darwin's finches and the less-famous mockingbirds (which Darwin called 'mocking-thrushes'). Altogether, Darwin wrote, 'several of the islands possess their own species of the tortoise, mocking-thrush, finches, and numerous plants, these species having the same general habits, occupying analogous situations, and obviously filling the same place in the natural economy of this archipelago, that strikes me with wonder'. And: 'we seem to be brought somewhat near to that great fact—that mystery of mysteries—the first appearance of new beings on this earth'.

When he wrote these words, in the 1840s, Darwin had an insight that in later years would prove to be very important in understanding how new species may come about—by geographicalal isolation. But Darwin never really put his heart into this idea. Instead, he later thought that new species could just as well be formed without isolation, by natural selection alone, a subject that we will return to later in this book. He left his earlier hunch to be picked up by some of his followers.

One of those followers was the German naturalist and explorer Moritz Wagner, who at the same time that Darwin was circumnavigating the world on the *Beagle*, had been travelling in Asia, Africa, and America. In 1837, while collecting insects in Algeria, the young Wagner was struck by the fact that each time he crossed a river, he would find new species of the wingless, blundering darkling beetle *Pimelia*. Just like Darwin in the Galápagos Islands, Wagner had found an example of species that were separated from their relatives by geographical barriers (rivers in Algeria that wingless beetles could not cross, stretches of sea in the Galápagos that 300-kilo tortoises would not swim).

Wagner seized on this idea and in 1868 he wrote a book about the subject, in which he proposed that 'the formation of . . . incipient species can succeed in nature only where some individuals can cross the previous borders of their range and segregate themselves for a long period from the other members of their species'. Wagner envisaged that a group of animals or plants would become cut off by some geographical barrier and start following their own evolutionary path: if the expatriates evolved a new characteristic, this would not be able to pass into the ancestral population, and vice versa. During a long period of separation, enough evolutionary changes could accumulate on either side of the barrier to transform the two populations into different species.

Wagner corresponded with Darwin on the subject, but the Englishman was not impressed. In a later edition of *On the Origin of Species*, Darwin wrote: 'I can by no means agree with this naturalist, that migration and isolation are necessary elements for the formation of new species'. And, in 1875, he privately brandished a recent paper by Wagner as 'Most Wretched Rubbish'. But Wagner was undaunted and towards the end of his life he expanded his theory further. In 1889, it appeared posthumously as *Die Entstehung der Arten durch räumliche Sonderung* ('The Origin of Species by Spatial Separation'). Hidden under this obscure German title, the theory lay dormant for more than half a century, until it was revived in the 1940s by the legendary Ernst Mayr.

The dew from heaven

On 6 September 1522, a ship named *Vittoria* sailed into one of the major ports of Spain, having completed the first-ever round trip of the globe. It was the single surviving vessel of the ill-fated fleet that had set out under Ferdinand Magellan years earlier. On board were masses of valuable and mysterious products from faraway places. Nutmeg, cloves, and other valuable spices, precious stones, and also two stuffed birds, a present from the Rajah of Bachian (ruler of the island of Tidore in the Moluccas) to the King of Spain. This may seem a meagre gift even by sixteenth-century standards, but what birds they were!

Nothing like them had ever been seen in Europe. The plumage was a dazzling palette of fiery red, bright chestnut, yellow, deep green, and iridescent yellowish green, completed with two tufts of amazing yellow-and-fawn, long, springy feathers. Even stranger, the birds had small heads, and no feet at all. Somehow, this particular bit of the ship's cargo attracted a lot of attention and soon the birds figured in many natural history accounts. In 1551, the Italian mathematician Girolamo Cardano wrote: 'Since they lack feet, they are obliged to fly continuously and live therefore in the highest sky far above the range of human vision . . . They require no other food or drink than dew from Heaven.' Nowadays, the birds' name still reflect that old legend: we know them as birds of paradise.

Of course, the 40 or so members of the family Paradisaeidae are no more celestial than any other bird. They are related to crows, but unlike those jet-black squatters of rubbish heaps, birds of paradise, which live only in the region of New Guinea and northern Australia, possess feathers in the most beautiful hues and forms. For this reason, they have always been sought after and used by Papuas for decoration and trading. And, of course, they do not lack feet and wings, nor do they have extraordinarily small heads. All this is just a consequence of the way the natives used to dry them.

The nineteenth-century English naturalist Alfred Russel Wallace, who travelled widely in New Guinea, gives a vivid description of such a procedure:

> [They] cut off the wings and feet, and then skin the body up to the beak, taking out the skull. A stout stick is then run up through the specimen coming out at the mouth. Round this some leaves are stuffed, and the whole is wrapped up in a palm spathe and dried in the smoky hut. By this plan the head, which is really large, is shrunk up almost to nothing, the body is much reduced and shortened, and the greatest prominence is given to the flowing plumage. [This gives] a most erroneous idea of the proportions of the living bird.

Nevertheless, the legend of birds of paradise is perpetuated in the official scientific name of the species that landed in Spain in 1522. Linnaeus called it *Paradisaea apoda* (literally, the legless bird of paradise) as a taxonomic quip. And the species is one of the pillars on which Mayr's full-fledged theory of geographical speciation is based.

It all began in 1927, when the young Ernst Mayr went to the International Zoological Congress in Budapest. There, he was introduced to Lord Walter Rothschild, patriarch of the rich and famous banking family. Apart from being a successful banker, Rothschild was also a passionate ornithologist, who owned a private museum at the family's estate in Tring, England. In addition to his museum staff, Rothschild employed hundreds of collectors all over the world. And when he met Mayr, he hired the young German immediately as his new bird collector to be based in New Guinea. For three years, Mayr explored the *terra incognita* of the New Guinean hinterland. Rothschild, who had a special affection for birds of paradise, had impressed on Mayr that he should collect as many specimens from as many localities as he could. That way, it would be possible to understand the exact distribution and variability of the many species. For birds of paradise, this is no trivial undertaking. Because of their beautiful colours, and the fact that literally millions of specimens had been imported into Europe to be turned into ladies' headgear, enthusiasm for birds of paradise had mushroomed everywhere among naturalists. And by the early twentieth century, hundreds of specific names were in use, with hardly anybody knowing whether a name referred to a species, a hybrid, or just a local variety.

Mayr put a lot of the confusion right. For example, he discovered that the dozen species of *Paradisaea* were all just plumage variants of a single species that gradually intergraded along the New Guinean coast. But he also noticed that the strongest differences occurred in areas that were particularly isolated, such as remote mountain ranges or islands. For example, the red bird of paradise, which is closely related to *Paradisaea apoda*, lives on the island of Waigeo, isolated from New Guinea's mainland by a strait 100 kilometres wide. The male differs from *apoda* by having red plumes, two long black streamers, and green tufts on its head, and so this *Uranornis ruber*, as the bird was known, was originally considered to be not just a different species (as the second part of the scientific name indicates) but even a separate genus (the first part of the scientific name).

Mayr discovered more examples of birds of paradise of which the most remote and isolated populations were also the most different in plumage, and which previous ornithologists had mistakenly placed into altogether different genera. For example, he recognized that what till then had been known as *Astrarchia stephaniae* from the remote Bismarck and Owen Stanley mountain ranges was merely a downgraded version of *Astrapia nigra*, the long-tailed bird of paradise.

Although his evolutionary ideas germinated during those New Guinean field trips, Mayr's career was only just starting. After collecting 3400 New Guinean bird skins for Lord Rothschild, he next became the leader on an expedition to the South Sea Islands, financed by the American millionaire philanthropist Harry Payne Whitney. Then, in 1931, he was hired by the American Museum of Natural History in New York to write up a bundle of publications describing the birds discovered on the Whitney expedition. The year after, Mayr went back to Rothschild's museum at Tring to take over the position of curator. But he had scarcely arrived in England when Rothschild unexpectedly announced that financial difficulties (blackmailing by a former mistress, as it later transpired) had forced him to put the entire 280 000-specimen bird collection up for sale. After some negotiations, the American Museum of Natural History finally bought it for the bargain price of $225 000 and Mayr returned, with 185 crates, back to New York.

It was in those stormy years that Mayr started to develop his theory of geographical or, as he called it, allopatric speciation (the term deriving from the Greek words *allos*, meaning 'other', and *patra*, or 'fatherland'). He had already noticed the striking influence of isolation on the New-Guinean birds of paradise. And, having amassed enormous materials and knowledge of the avifauna of the Malay archipelago and the South Sea Islands, he soon began to see proof of allopatric speciation in scores of other bird groups, ranging from robins to kingfishers.

Initially, Mayr stuck to his feathered friends to find support for his theory, but he soon started to look for further examples outside the realm of birds. Even in the days before geologists had fully realized that continents actually drift, crack, break, and fold, Mayr understood that, on a geological timescale, any animals' environment is continuously changing: rivers meander, mountain ranges emerge, seas overflow coastal areas to form islands, volcanoes erupt, ice ages spread glacial blankets, and so on. Earth's forces are forever splitting up animal ranges, producing isolated populations everywhere all the time. And animals themselves may disperse to faraway places and become cut off from their relatives.

The longer a period of isolation persisted, the more genes would change by the steady rain of mutation, Mayr thought. Random hereditary errors would occur here and there, and the accumulated mutations would gradually change the genetic make-up of the population. Depending on how long a period of isolation persisted, such isolated populations could either not change at all, change a little, or evolve into altogether new species, which would no longer be able to interbreed. In the latter case, whenever the geological barrier lifted, two species would have been formed where there originally was only one. And if the animals' ecological niches had also evolved, the animals could gradually spread into each other's home ranges without merging genetically: the two would become sympatric (from *sym* meaning the same, and again, *patra*, the fatherland).

It was a plausible enough theory, which immediately won a lot of support. But how was Mayr to prove that this was the way species actually formed? Like so many evolutionary processes, Mayr realized that allopatric speciation would take thousands if not millions of years, and could never be seen happening from start to finish. So he used an indirect approach. In 1963, his book *Animal Species and Evolution* appeared, a *tour de force* in which Mayr integrated a vast amount of information from all corners of biology into a modern view of evolution and speciation. In it, he wrote: 'Any scientist who must interpret past events, like the archeologist, the historian, or the geologist, or who, like the cytologist, studies dynamic processes that can be observed only at definite fixed stages . . . solves it with the same method.' Mayr had decided that if we were to discover whether allopatry was the true mechanism underlying speciation, we had to look in nature for all possible intermediate points in the speciation process and then mentally join the dots to get a picture.

In the next four short stories, we will follow Mayr's example. Each of the tales may represent an intermediate stage in allopatric speciation. Of course, we can never be certain that these individual cases will in future lead to full species (many of them probably won't), nor can we regard it as hard evidence that speciation always proceeds through similar phases. Nevertheless, if allopatric speciation is going on in nature, we would expect to find all these kinds of intermediate stages.

The giant wedge

Long ago—in fact, around 90 million years ago—the chunk of land that was later to become South America split off from Australia and Antarctica. For almost the rest of its geological life, the island-continent drifted aimlessly across the globe. Then, roughly 3 million years ago, some massive plate-crunching uplifted a stretch of land that precisely bridged the gap between South America and its neighbour North America. The thin land-bridge, which we now know as the Isthmus of Panama, had a dramatic impact on the fauna of the region. Most famously, it set in motion an episode affectionately known to palaeontologists as GAFI, or the Great American Faunal Interchange.

South America, when it split from Australia, had taken with it its share of marsupial mammals. North America, on the other hand, had played its part in the evolution of modern mammals, so it boasted a fair collection of these. When the Panama Isthmus rather suddenly reconnected these two different faunas, a steady flow of dogs, cats, and camels travelled into South America, while opossums, armadillos, and other primitive mammals went the other way: the faunal interchange. It is less well known that in the sea, the reverse took place. Before 3 million years ago, the seaway in between the two continents had been a common ground for a rich variety of marine creatures. But the isthmus changed all that: it split all those populations in two, creating ideal circumstances for allopatric speciation.

'It is a sort of textbook case', says Nancy Knowlton, a marine biologist based at the Scripps Institute of Oceanography in La

Jolla, California. For most of the 1990s, Knowlton was studying a group of animals that have behaved exactly as Ernst Mayr predicted since the isthmus came into place. Her favourites are called snapping shrimps. They are between 1 and 5 centimetres long, and they sport one enlarged claw with which they strike blows to defend themselves. They can also snap shut the small finger at the end of the claw, which creates a loud clicking sound. Hence their vernacular name of pistol shrimps. 'These are pretty aggressive little creatures', says Knowlton respectfully. 'They'll snap at anything that's not a potential mate, and actually if you left a few in a dish of water, without any places to hide or escape, then probably one of them would fairly quickly kill the other one.'

Despite their violent nature, snapping shrimps are pleasantly compliant where allopatric speciation is concerned, as Knowlton found out in the early 1990s, when she spent years collecting the animals on both sides of the isthmus. Among her bounty were seven Pacific species that, from the way they looked, seemed to have Caribbean counterparts. When Knowlton identified the shrimps, she found that some trans-isthmus twins are clearly the same species whereas others are so different that earlier biologists had given them different names. But whether they have the same name or not, in all the pairs she noticed subtle differences in, for example, their colour patterns.

The genes of the shrimp twins are different, too. Knowlton singled out a chunk of DNA and studied the sequences. Sure enough, the DNA in each twin was similar but not identical. She found between 8 and 20% DNA difference between the two members (1% difference means that one out of every 100 of the 'DNA letters'—A, C, G, or T—has mutated to another letter). The DNA differences mean that each species has, over time, accumulated some random mutations across the isthmus. The longer they have been separated, the more genetic differences there are. This fits nicely with the habitats where the shrimp species live. The pairs with the largest genetic difference live further out in the deep water, whereas the ones from mangroves and other coastal habitats are the least diverged genetically. This is easily explained: the species living in shallow water would have been the last to be separated by the rising land-bridge. Given that the rate of mutation in this particular piece of DNA is known (about 2% per million years), Knowlton had found a pretty exact fit with the region's geological history. Most of her Pacific–Caribbean shrimp pairs had started changing genetically between 3 and 4 million years ago: the precise time the isthmus rose from the sea and isolated the populations.

The shrimps' intolerance towards one another proved a useful asset when Knowlton had a closer look at yet another characteristic: reproductive isolation. Having not been in contact for at least 3 million years, were the Pacific and Caribbean representatives of a species still able to recognize each other as potential mates? Many of them were not. When Knowlton did tests by putting males and females from either side of the isthmus together in a bowl of sea water, they approached each other with the hostility they usually reserved for members of the same sex or for different species. Only one out of over 50 transisthmian couples managed to hit it off and produce a clutch of eggs, whereas males and females of the same species from the same ocean got along swimmingly most of the time.

'My take on the Isthmus is that for many, many groups 3 million years is long enough to create sufficient divergence that reproduction is no longer likely', says Knowlton, who meanwhile has discovered an additional eight pairs of transisthmian snapping shrimps. That may be so, but less time and a much more modest setting can provoke geographical speciation just as well, as it has in insects from the other end of the Atlantic.

Chasms cause schisms

Near the town of St Girons, on the north-facing flanks of the Pyrenees, lies the French Underground Laboratory. Its name derives not from any sinister or illegal activities that go on there, it is just literally underground. In the 1940s the French national scientific research organization (CNRS) built there, half in a cavern, half outside it, a laboratory to study the ecology of caves.

Biologists have always had a strange fascination with caves. The speleologist's sense of adventure to explore vast underground halls and passageways where new and unexpected life forms may lurk, obviously plays a part. But there is also a more rational reason why caverns are biologically interesting. The environment is radically different from the one above ground. In caves, there are no seasons. It is always humid and cool. There is little food: all the nutrients are either brought in by bats or by percolating ground water. Above all, there is no light. And in evolution, such harsh conditions call for drastic measures.

Caves are theatres of evolutionary adaptation. Insects and other small animals that somehow penetrated the deep recesses of caves have, over thousands of generations, been changed into sometimes quite bizarre creatures. For example, a beetle called *Leptodirus hohenwarti* has been moulded into something more resembling a Giacometti sculpture in its cave home in Croatia. Incredibly thin, brittle legs carry its pale balloon-shaped body, from which springs a narrow eyeless face with long, threadlike antennae. The long legs allow the animal to cover vast distances

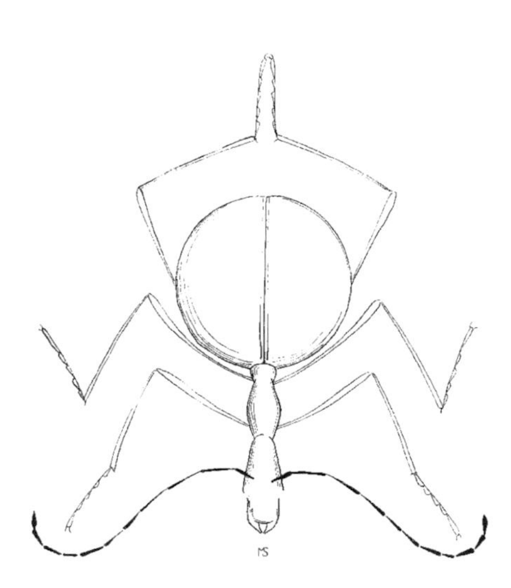

Cave beetle, *Leptodirus hohenwarti*, found in Croatia. Its large body holds an extended digestive system to make the best of the limited food available.

because food sources are few and far between. Its large body holds an extended digestive system that makes the best of what little food there is. And about the lack of eyes in blind troglodytes like this one Darwin wrote: 'Disuse will . . . have more or less perfectly obliterated its eyes, and natural selection will often have effected other changes, such as an increase in the length of the antennae or palpi, as a compensation for blindness.'

Members of the beetle group to which this intraterrestrial being belongs live in caverns all over southern Europe, presumably the descendants of soil-dwelling ancestors that have long since gone extinct. In the laboratory in St Girons, Christian Juberthie and his wife Lysiane Juberthie-Jupeau have been studying another member of this group, *Speonomus brucki*. Although this beetle is not so weird in general appearance as the Croatian one, it is still blind and has lost all its colour. *Speonomus brucki* is only 3 millimetres long but inhabits numerous different Pyrenean caves in a region 60 by 100 kilometres.

Until 1980, the French team were puzzled by this phenomenon. They knew from experiments in their subterranean lab that *Speonomus* beetles are strictly dependent on a cave environment and certainly would not be able to live outside it. So how on earth (or, rather, how under the earth) did they manage to spread from one cave to the next? In 1980, Christian Juberthie discovered that, for these tiny insects, many more spaces qualify as 'caves' than just the large, air-filled grottoes that we humans call by that name. He and his colleagues started digging into the rocks that lay under thick layers of forest soil and found *Speonomus* beetles thriving in the centimetre-sized cracks and fissures that were produced in the deep bedrock during the ice ages of the past 1 million years. These newly discovered microcaverns, which are abundant passageways in between the more spacious caves, offered a good explanation for the widespread occurrence of the beetles. But this did not mean that all *Speonomus* beetles in the area were identical. When Lysiane Juberthie-Jupeau looked at the mating behaviour of the beetles, she discovered striking differences between the beetle groups.

A series of caves and microcaverns at an altitude of roughly 900 metres all harbour beetles with the same sort of mating behaviour. Juberthie-Jupeau observed that a male would hop onto a female, insert his penis, and move it about vigorously. Then it would give her abdomen some rest by tapping with his antennae on her back instead. Next came a phase with a combination of penis thrusts and antennal taps, followed by a short period when the male would be motionless before he dismounted. The whole affair would be over in less than 4 minutes.

Just a few kilometres downhill, at an altitude of about 700 metres, Juberthie-Jupeau discovered a cavern near the village of Banat where the cave beetles did things differently. The copulation lasted twice as long. And instead of the monotonous up and down strokes of the penis, the males of this population varied their performance. First came a burst of strong and slow thrusts, then a flurry of gentle and quick ones.

And some 12 kilometres to the north, in the Crouanques Cave, 500 metres above sea level, Juberthie-Jupeau discovered yet another type of mating behaviour in *Speonomus brucki*. Here, the male would sit on top of the female with his penis inside her, but not moving his body at all. Only his antennae would tap the female all the time, and now and then he would throw in some flank-rubbing with his legs as well. The entire copulation would go on uninterruptedly for a full 15 minutes.

Juberthie-Jupeau and her husband then decided to have a look at the genetic differences between the various beetle populations. They did this with a technique called protein electrophoresis. Until DNA methodology took over in the 1990s, this was the standard way to get an impression of the genetic difference between two populations of organisms. It is less sophisticated than it sounds. Basically, it involves putting squashed bits of the animals on one end of a slab of gel, then letting a strong electric current pull the proteins through it, after which the gel is 'developed' in a bath that stains the proteins. Genetic differences produce different proteins, which, in turn, are pulled faster or slower through the gel by the electric current. So if two individuals have 15 bands at the same positions, but one or two at different positions in the gel, this means they are genetically rather similar. If all their bands are at different positions, they must be genetically very different.

The Juperthies did this with beetles from four places along the 900-metre contour, and those variant ones from Banat and Crouanques. They found that the Banat and Crouanques populations both stood quite apart genetically, whereas the other four were genetically similar. This matched Juberthie-Jupeau's observations of their sexual behaviour. But how had this come about? In 1988, the researchers published their answer: the Ice Ages. Nowadays a belt of forest lies between 800 and 2000 metres altitude. Above it lies alpine tundra and below is dry Mediterranean shrub land. During the coldest snaps in the Pleistocene, the forest belt descended to the foot of the mountains, reaching down to just a few hundred metres above sea level. The Juberthies knew this from archaeological studies of pollen in the soil.

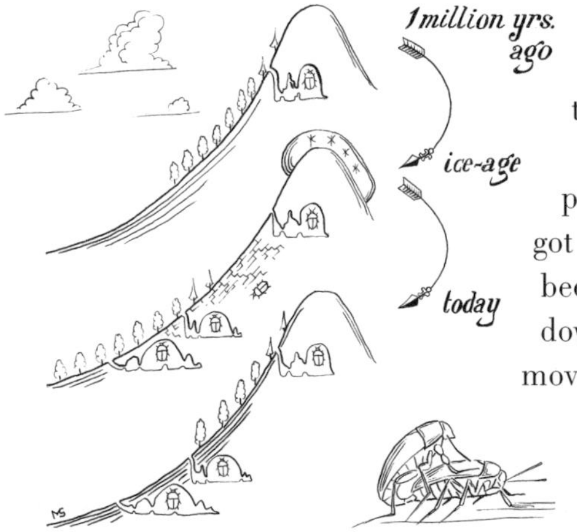

What must have happened as the climate got colder is that the cave beetles slowly crept downhill with the forest, moving through cracks and microcaverns until they reached the area below 500 metres. When the climate improved again, the trees and

The sequence of events that led to the geographical isolation of populations of the cave beetle *Speonomus brucki*.

most of the beetles moved back up the slopes. But the deep freeze and its aftermath, meltwater, left new caves at the lower altitudes, and these became the home of some populations of *Speonomus brucki*. These cave beetles stayed behind when the forests retreated up the mountain. Next, erosion took away the soil and sediment filled up any cracks in the rocks on the slopes between the present-day lower and higher populations, so that these became isolated from each other. On the basis of the protein differences, Christian Juberthie estimates that the separation happened less than a million years ago. Since isolation, little by little the beetle populations in Crouanques and Banat changed their sexual behaviour, even though their other habits and appearance remain almost identical. Nevertheless, such little shifts in behaviour can eventually lead to speciation, as is attested by the famous Ring of Gulls.

A ring of gulls

Martin Reid's web site has a teaser. You get to see four pictures of almost identical-looking seagulls. Three of them are Siberian gulls, the fourth is something else. And Martin wants you to e-mail him which of the quartet is the odd one out, what it is, and why. 'Note: I have distorted any other gulls in the background, to remove clues', he adds helpfully. The puzzle will be familiar to die-hard birders. Because on hearing the 'gah-gah-gah' of herring gulls and their relatives, a birdwatcher will either look up in anticipation or look away in disgust, depending on whether or not he or she feels like the tedious chore of identifying these birds.

Over a dozen subspecies of this type of gull exist in the Northern Hemisphere. Each differs from the next by tiny variations in the colour of their feet, the shade of grey on their wings, the size of their beaks, or the colour of their eyes. In 1934, the ornithologist B. Stegmann realized that many of these forms are links in a chain that forms a ring around the Arctic Circle. The lesser black-backed gull of western Europe freely interbreeds with the Baltic gull of northern Europe. This one, in turn, passes into the Siberian gull, and so on all the way around the globe, via the Vega gull of eastern Siberia, the American herring gull and the herring gull of Europe. But wait. Did we not start the round tour in Europe? Yes we did, and that is precisely why the herring gull example is so special. The pink-legged, grey-winged herring gull and the yellow-legged, black-winged lesser black-backed gull form the two ends of the chain and, where they meet, they behave as completely separate species.

During the 1950s, several biologists had a closer look at the seagull enigma. They superimposed the present-day home ranges of the various gull subspecies on maps with the glacier distributions during the cool spells of the Pleistocene, and found that the ones in Siberia must have been isolated in small southerly 'refuges' during the cold times. The various subspecies apparently evolved in isolation there. During successive glaciations, new subspecies split off in more easterly refuges, producing a whole chain of forms, each a little bit different from the previous one but not sufficiently different to prevent them from interbreeding when warm periods (like the present) brought their breeding grounds in contact again.

After the end of the last ice age, around 10 000 years ago, the latest offshoot of this budding sequence, the herring gull, crossed the Atlantic and there, finally, was reunited with its long-lost ancestor, the lesser black-backed gull. But after so many thousands of years of not having been in touch, the two had drifted apart in lots of ways. In 1955, the German birder Friedrich Goethe put together an exhaustive enumeration of all the differences he could find. Here is a selection: herring gulls do not migrate, live near inland and shore waters, are beach-combers, are more tame and during copulation utter the sound 'agchagcha'; lesser black-backed gulls, on the other hand, migrate, live on the open sea, catch fish for a living, are not so tame, breed two or three weeks later than herring gulls, and, while mating, gurgle 'go(a)go(a)', 'gäägäägääg', or 'gräägrää-gräää g'. No wonder that in the occasional mixed colony, the lesser black-backed are 'not taken seriously' by the herring gulls, as Goethe observed.

To Ernst Mayr, ring species like this one were living proof of allopatric speciation. Slightly different subspecies, each formed by geographical isolation, and each a little different from the next, were lined up in a circle, leading eventually to mature species at the termini. All the stages of allopatric speciation were there, captured as if in a time capsule. In 1940, Mayr wrote: 'There is no better way than this to demonstrate the effectiveness of geographical variation.' Until, that is, Humphry Greenwood and his amazing fish came along.

Fish on the edge

Africa is being broken up by relentless geological forces that are pulling at its eastern shore. Millions of years from today, the land mass that now hosts the countries of Tanzania, Kenya, Ethiopia, and Somalia will be a Madagascar-like island somewhere in the middle of the Indian Ocean. Already a string of lakes (Lake Malawi, Lake Tanganyika, and Lake Victoria, to name the three largest ones) betrays the seams at which the great continent is coming apart. At some point in the future, the sea will enter the Rift Valley and flush out all the fresh water from the three great lakes, extinguishing forever what must have been one of the most spectacular evolutionary explosions: the cichlid species flock.

Between them, the three lakes hold an amazing 2000 different species of the fish family Cichlidae—or *furu*, *mbuna*, and *mbipi*, as the local fishermen call them. And since their discovery, zoologists (not fishermen) have marvelled at their staggering diversity, not only in colours or looks but also in their modes of life. The cichlid species flock is no less than an underwater counterpart to Darwin's finches, amplified a hundred times.

Starting from only a few ancestors, the cichlids have fanned out to specialize on every type of food conceivable, changing the shapes of their bodies to suit their lifestyle. 'It's hell on wheels', cichlid specialist Les Kaufman recently exclaimed on an e-mail discussion list.

Lake Victoria, for example, houses robust, heavy-jawed snail-crushers, slender and swift plankton-eaters, small cleaner-fish that peck parasites off other fishes' skins, big-mouthed, mean-looking fish-eaters, and stout, sharp-toothed algae-scrapers, to name but a few of the many trades that these perch-like fish have taken up. Setting new standards for ecological specialization, some species from Lake Tanganyika scrape the scales from only the right flanks of other cichlids, while other species take care of the left flanks.

Lake Victoria is not an old lake. Until recently, it was thought to be only half a million years old and, as we shall see in Chapter 9, even that may be an overestimate. So where did the 500 or so cichlid species that live there all come from? At first, some people suggested they immigrated from surrounding rivers, but molecular biologist Axel Meyer showed in 1990 that they all have very similar DNA, so they cannot derive from different sources. Instead, they must have all evolved in the lake itself. So then people thought that perhaps there were unseen barriers in the lake that had split up the fish into different species. But nobody has ever found such barriers. In fact, the lake bed is shaped like an almost perfect saucer, with a maximum depth of less than 80 metres.

In 1965, Humphry Greenwood, the fish curator at the British Museum (Natural History)—as London's Natural History Museum was then called—came up with an answer. A few years earlier, a group of Cambridge students had brought him cichlids they had collected in Lake Nabugabo, in Uganda. This 10-kilometre-wide lake, little more than a swamp at the edge of Lake Victoria, is separated from its big sister by only an unimpressive sandy ridge of a few hundred metres across. Yet, among the fish samples the students brought back, Greenwood discovered no less than five new and unique cichlid species.

Two of them were plankton-eaters, one was a fish-eater, and the remaining two were vegetarians. For each, Greenwood could point at a Victoria counterpart. Nevertheless, the Nabugabo species differed from their Victoria sister species by the shape and numbers of their teeth, and, most strikingly, by the coloration of the males—always a spectacular sight in cichlids from the African great lakes. For example, Greenwood described *Haplochromis simpsoni* from Nabugabo as follows:

> Dorsal surface of body cobalt, that of the head dark umber to black; flanks with light turquoise sheen, the ventral body surface sooty except for the chest which is silvery-white with a diffuse sooty overlay.

Its close relative *H. empodisma* from Lake Victoria, Greenwood wrote, differs by having a characteristically red head. There was no doubt that the five new forms were species in their own right. 'Indeed,' Greenwood commented, 'the morphological gaps between any one Nabugabo species and its counterpart in Victoria are as great as those between any two related *Haplochromis* in Lake Victoria.'

In all probability, the five Victorian ancestors of these five species had become cut off when the water level of Lake Victoria dropped a few metres, and Lake Nabugabo was formed. Here, they had evolved their own nuptial coloration and the other characteristics. The fact that the border between the two lakes is only a narrow sandbar suggests that the event happened not too long ago. And, indeed, on the far side of Lake Nabugabo, a team of British geologists have discovered the traces of what must have been a former Lake Victoria coastline, lying above the height of the sandbar. Some chunks of charcoal were found here, rolled by the waves at the long-gone shore, which the geologists dated with the radioactive carbon method at only 4000 years old.

So five new species had managed to evolve in the (evolutionarily speaking) short span of time that elapsed since the pharaohs ruled Egypt. And Lake Nabugabo is not alone. The 24 000 kilometres of sinuous Victoria shoreline is dotted with many such tiny satellite lakes and during the lake's lifetime they probably have come and gone many times over. So if 4000 years is enough time for five new species to evolve in one of these lakes, the entire Victoria species flock could easily have been generated in the many satellite lakes that must have existed over the many thousands of years since the lake was born. 'It could hardly be more pertinent to the problem of cichlid speciation in Lake Victoria', Greenwood wrote.

Greenwood's work is equally pertinent to the theory of allopatric speciation as a whole. For if we mentally join this latter dot with the three from the previous sections, the picture we get is allopatric speciation as Mayr saw it. Geographical isolation breeds genetic differences (Panama snapping shrimps). These can eventually shine through in sexual behaviour (beetles), and colour and form (gulls). And, when stretched far enough, separate species evolve that can live in the same place thanks to their different ecological niches and because they no longer interbreed (gull ring species). In fact, the whole sequence of events can be telescoped in very small populations on the rim of a species' range (Nabugabo cichlids).

CHAPTER THREE

Tight Spots

The Magic of Small Populations

Ernst Mayr was particularly fond of the spectacular sort of rapid allopatric speciation that had apparently gone on in the Lake Nabugabo cichlids, at the edge of Lake Victoria. He called it the 'peripheral isolates' model. When Mayr was on expeditions in South-East Asia and Oceania, he had time and time again noticed that the birds on small islands or tiny isolated populations on the rim of a species' range, the periphery, were usually the most deviant in looks. Apparently, a different sort of evolution went on in those tight spots.

Take, for instance, the curiously named spangled drongo, a member of the crow lineage from the Malay archipelago. While all the birds in the central populations from Indonesia are rather uniformly black with slightly forked tails, the ones from places at the very edges of the range, such as New Ireland and the Philippines, have exaggerated tail feathers that are long, wavy, and curved inwards or out. Another example that Mayr exhaustively used in his writings was the paradise kingfisher. On the vast mainland of New Guinea this kingfisher looks pretty much the same everywhere, but the ones on small islands off the New Guinean coast without exception have strikingly different plumages. 'Whenever we find a representative of this group on an island,' Mayr wrote, 'it is so different that five of the six Papuan island forms were described as separate species and four are still so regarded.'

In 1942, in his influential book *Systematics and the Origin of Species*, Mayr offered an explanation for this curious peripheral

Ernst Mayr's peripheral isolates: the marginal drongo populations are different.

isolates effect. Often on these far-flung spots, Mayr reasoned, 'the entire population was started by a single pair or by a single fertilized female. These 'founders' carried with them only a very small proportion of the variability of the parent population'.

What Mayr had in mind is something now known as the 'founder' or genetic 'bottleneck' effect. Think of a handful of coins scattered on the floor. If you have a hundred coins, the chance of all of them lying heads up is virtually zero. But with only a few coins, say five or six, the chances of this happening are fair. In genetics, it is no different. Imagine a far-away place colonized by three individuals from a large population. In the ancestral population are three alleles for a certain skin-colour gene: let's call them red, yellow, and blue. Red and yellow are common, with frequencies of 50% and 45%, respectively; blue is much rarer, with a frequency of only 5%. Two of the three colonists have only red alleles, while the third has a red and a blue allele. No yellow alleles happened to come along. After the founder event, the new population has very different frequencies of these alleles. By coincidence, red has gone up from 50 to 83%, blue from 5 to 17%, while yellow has completely disappeared. If the population remains small, a series of such accidental genetic turnabouts could happen. The whole process is called 'genetic drift', and we will return to it again.

Bottlenecks and the consequent genetic drift play a part in human populations, too. A certain remote village in the Alps has a high prevalence of albinism. And on the island of Tristan da Cunha, stocked by a single Scots family, a rare hereditary eye disease is rife. Possibly the founding father of that family happened to carry on one of his chromosomes the faulty allele that causes this eye disease. As a result, half of his children (that is, 50% of the new population) inherited it. Mayr thought that such sudden changes in genetic make-up might also lead to speciation in peripheral isolates. During the 1940s he realized that he was on to something interesting here, and started nurturing his notion of bottleneck speciation. The time was right because these were days of great upheaval in biological science.

The great amalgamation

In the 1930s, three mathematically inclined biologists captured Darwin's theory of natural selection in solid mathematical equations. They were the Englishmen Ronald A. Fisher and J. B. S. Haldane, and the American Sewall Wright. While they were at it, they also included other processes that could change the genetic make-up of populations, such as mutation, gene flow, inbreeding, and things like genetic bottlenecks. By the time Mayr returned from his ornithological expeditions and settled in the American Museum of Natural History, so-called 'population genetics' was a well-founded discipline, enabling biologists to study evolution in the same rigorous way that physics and chemistry was done: with numbers, equations, and predictions.

One of the first biologists to adopt the new faith was Theodosius Dobzhansky. In 1927, this Russian naturalist had come to New York's Columbia University to continue his work on the genetics of banana flies—commonly, but erroneously, also called fruit flies. With these tiny and prolific insects that have a life cycle of less than two weeks, Dobzhansky provided evolutionary research with yet another asset: the ability to study it in the laboratory. In the bottles and jars with fermenting fruits in Columbia's 'fly room', Dobzhansky watched evolution happening.

In 1936, Mayr invited Dobzhansky over to the museum to have a look at his bird skins and listen to his ideas about allopatric speciation. Dobzhansky was impressed and, like Mayr, embraced geographical isolation as the principal means by which new species started out. In later years, he even required his students to write down the definition of allopatric speciation for their entrance examination at Columbia. The next year, when he was forced to stay in bed for weeks after a horse-riding accident, Dobzhansky wrote his book *Genetics and the Origin of Species*, in which he amalgamated genetics, population genetics, and natural history into a modern, broad, and understandable view of evolution. It was to be the manifesto for the revolution that, as Dobzhansky later said, 'was in the air'.

Over the next ten years or so, everybody who was anybody in biology jumped on the bandwagon. In *Systematics and the Origin of Species*, Mayr upgraded his taxonomical studies from pigeon-holing to an enterprise that could really contribute to understanding evolution. His fossil-studying colleague George Gaylord Simpson did the same for palaeontology two years later, and Julian Huxley, the grandson of Darwin's friend and intellectual bodyguard Thomas Henry Huxley, wrapped it all up in his book *Evolution: The Modern Synthesis*. It was during those exciting years of the Modern Synthesis that Mayr began to develop his bottleneck model. Somehow, the founder effect on its own did not seem strong enough. So, drawing heavily on the wealth of genetic insights that had become public property during the Modern Synthesis, he sculpted his 1942 preliminary sketch into a monumental theory that was published in 1954 as a paper entitled 'Change of genetic environment and evolution'.

The creative force of a marginal existence

To start with, Mayr knew that animals are not just 'piles of genes'. Instead, they are integrated, balanced networks, where no gene is an island by itself. Take this hypothetical example: an animal has a gene that makes its tail bright red. Nevertheless, the gene will work only if the animal also has genes that make a tail. This may sound like a silly truism, but it is how the genetic environment works. The way an organism looks and functions is not just a simple addition of all its genes: it is the product of complicated networks of gene–gene interactions. More realistically, a person can carry all three genes that induce a dark skin, and yet be a complete albino if the gene for tyrosinase, which produces the melanin needed for dark coloration is missing.

These interactions make natural selection work in unpredictable ways. The hypothetical red-tail gene can be selected in a population only where there are tails that need colouring. Natural populations are, of course, variable. So, Mayr envisaged a situation where most of the individuals have no tail. Hence if a gene appears that makes tails red, it will probably not be able to function, given that it has few tails to work on. But if, due to chance, a group of colonists, among which there are a lot of tails, arrives on an island, the red-tail gene will spread much more easily in this new genetic environment.

These gene–gene interactions are little understood, even today. They may be one-on-one, but may also be composed of interminable chains and networks of mutual influence. So bottlenecks can have huge effects on the genetic make-up of a population arising from founders. Rare genes can suddenly become common whereas common ones might disappear. A gene that in the ancestral population had led a slumbering existence might suddenly be awakened by the appearance of some other gene that required it. The ensuing cascade, Mayr wrote in his 1954 paper, 'may have the character of a veritable "genetic revolution"'. Mayr saw that the whole process could snowball out of control, leading to a new integrated complex of genes. He later developed an entire vocabulary of catchy terms to describe the process. In his words, the 'chain reaction' would lead to a 'genetic reconstruction' or 'genetic shake-up', which would be no less than an 'evolutionary rejuvenation' and gave the new species a 'clean start'. All of this could not happen in the 'dead heart' of the species. Clearly, he was very fond of his baby, which he later said had been 'perhaps the most original theory I have ever proposed'.

Although Mayr admitted in the postscript to his benchmark paper that the whole thing was 'frankly speculative', the theory became immensely popular with his fellow evolutionists. It was the ideal mix of genetics, ecology, and taxonomy, marinaded in Darwin and topped with a good dash of biogeography, which made his model particularly palatable in the prevailing synthetic atmosphere. The model even offered a way out of the palaeontological conundrum of gaps in the fossil record. After all, if new species emerged by rapid evolution in small, isolated populations, it was no wonder that the transitional forms were never found. In the 1970s, palaeontologists Niles Eldredge and Stephen Jay Gould exhumed and revived this notion as the 'punctuated equilibria' theory.

But however appealing, the bottleneck model badly needed data. It needed real-life examples from nature and from the laboratory that the genomic overhauls that Mayr envisaged really could and did happen. It took some time, but eventually the data started streaming in from studies on banana flies and marine worms.

No, we have plenty of bananas

Most geneticists will view their lab banana flies only as handy gene machines that live in incubators. But *Drosophila* flies are wild animals, and *Drosophila melanogaster*, the species that is most often used in genetics research, is just one of 1500 such species living all over the world.

Little more than a few volcanic specks in the immense Pacific Ocean, the Hawaiian Islands have provided the stage for the pinnacle of banana-fly evolution. Begun in 1963, the scientists of the Hawaiian Drosophilidae Project have discovered about 500 different species there. Some 250 additional species still sit in drawers and vials awaiting formal description by taxonomists, while an estimated 300 are yet to be discovered on the more remote mountain slopes (altogether a whopping 1000 species on an area only a little larger than London).

In fact, these *Drosophila* species are strikingly different from the cloud of *melanogaster* that envelops your fruit bowl when your bananas are past their prime. They live in habitats as diverse as the Hawaiian Islands can offer, from sea level to the highest mountain tops. Some go for desert-like patches where they breed in decaying cactus leaves, whereas others live in rainforest on tropical fruits or fungi. One species even lays its eggs in spiders, while another has larvae that live under water. Many are large, housefly-sized things. In species with cheerfully mottled wings, the males perform dances in spots called leks (see Chapter 4) where the females watch them and mate with the ones they fancy. The males of one species curl their abdomens over their heads and shower potential mates with a cocktail of aphrodisiacs. And, making a bid for supremacy in banana-fly weirdness, *Drosophila heteroneura* males have hammer-shaped heads with eyes mounted on stalks, with which they engage in head-butting fights as if they were miniature elks.

This home-grown *Drosophila* species flock owes its existence to Hawaiian geological history. Under the sea off the easternmost and largest island, Hawaii, sits a stationary thermal plume. Over the past millions of years, this outlet of the liquid inner regions of Earth has been regularly spewing lava over the seabed. Because the seabed itself is not stationary, but moves in a westerly direction, the plume has produced a string of islands like a geological conveyor belt. As a result, the westernmost island of Kauai was formed about 5 million years ago, while Hawaii is only 500 000 years old. Even at this moment, a new island (provisionally called 'Loihi') is being heaped up underneath the sea 3 kilometres off Hawaii.

When geneticists started studying the DNA of the Hawaiian flies, setting up a family tree, they discovered that the flies' evolution exactly matched the geological history of the archipelago. The species that had evolved first, branching off at the root of the family tree, lived on the oldest islands, while the most recent offshoots occupied the youngest islands. In fact, even within the Big Island the same pattern reappeared, with the western part, which lies closest to the volcanic hotspot, harbouring younger species than the eastern half. So apparently, the flies had been island-hopping onto each new volcanic island, as soon as it cooled down, spawning new species all the time.

What has gone on among the Hawaiian banana flies is a series of thousands of founder events—not just the colonization of newly formed islands but also the colonization of islands within islands. Each of the Hawaiian Islands probably took tens of thousands of years to form, the erratic accumulation of lava from the whimsical thermal plume. Vegetation would build up on the lava flows, only to be bisected by more recent flows. As a result, tiny pockets of banana-fly habitat would come and go all the time, providing ample opportunity for the foundation of myriads of *Drosophila* colonies. Even today, many unique species occur in rainforest pockets in between lava flows that are so typical that the Hawaiians have a special word (*kipukas*) for them.

In 1968, Hampton Carson, one of the more prominent Hawaiian drosophilists, thought up his own adaptation of Mayr's peripheral isolates model. He realized that the founder event may indeed be important in the origin of new banana-fly species in *kipukas*. But he also saw an important role for the subsequent rapid proliferation. Suppose a pregnant female fly landed in a new *kipuka* where there were plenty of bananas but no banana flies to eat them. In less than no time, the fly's progeny would have filled the vacant niche.

Carson saw that there would not only be the opportunity for a Mayrian genetic revolution, but also that the flies would be free from natural selection during the population 'flush' that followed the founder event. After all, natural selection works only when there is insufficient food for everybody, so that only the fittest survive. But in Carson's scenario, for many generations there would be enough bananas to feed even the genetically least endowed. Genes and gene combinations that would never have made it in the overcrowded population that the grandmother-fly had hailed from could spread unchecked in the still-empty *kipuka*. So there were two steps to it: the first was the founder effect, and the second was the possible change in gene frequencies during the selection-free 'flush' phase. Carson's double act became known as the 'founder–flush' model.

The founder–flush proved a welcome addition to Mayr's traditional model. Because the flush phase provided additional scope for rapid evolution, and because flushes would happen whenever a few founders colonized a vacant niche, it could be widely applicable. Support for founder–flush and similar models of speciation came not only from wild banana flies: their relatives in incubators in geneticists' labs soon put in their pennyworth as well.

Bottles and their necks

In 1974, Hampton Carson's genetics student Alan Templeton was studying one particular Hawaiian banana fly, *Drosophila mercatorum*, which lives on decaying leaves of 'prickly pear' cactuses on the slopes of the Kohala volcano on Hawaii. Like most banana flies, this species has the latent ability to perform virgin birth, or parthenogenesis. A female can duplicate an egg cell to trick it into 'thinking' it has been fertilized. The egg cell, now with a double, identical set of chromosomes, then begins developing normally. On the dry slopes of the volcano, where rotting cactus pads are few and far between, parthenogenesis is a useful asset. After all, as Templeton once said, it is an easy way out if a female finds itself on a cactus pad with no male around to fertilize her eggs.

Parthenogenesis was also a useful quality when Templeton began laboratory experiments on the founder-flush model, using his cactus flies. He collected 34 female flies and let them lay eggs. Their virgin daughters he put by themselves in lab bottles. Some of these, believing they had landed on uninhabited cactus pads, started doing their parthenogenetic thing and produced offspring in the biblical fashion. By doing so, they squeezed through the ultimate genetic bottleneck: a single founder individual. As Templeton wrote, the reason for his approach was to push the essence of the founder–flush model to its absolute extreme. That way, he might speed up evolution and watch new species appear in the lab.

The offspring of these parthenogenetic females were genetically all different, because each had been born from a single, duplicated egg cell (as you'll recall, egg cells of one female are all different because of recombination). Templeton then took some of these daughters and put them once again by themselves in vials. Again, the females produced daughters without sex. But this time—being produced by a single egg cell—these mothers had the same allele at both copies of each of their genes, so their daughters were all clones of herself. This gave Templeton a lot of genetically identical starting material for his experiments.

He then confined these clonal daughters individually in vials, added a male or two from another stock, and sat down to see what would happen. In one stock, named K28–0-Im, he noticed something peculiar. Females from this stock were very reluctant to mate. In a total of over 200 trials, he saw only eight copulations. And out of the sons that were born from those rare matings, a quarter were sterile, having empty testes. As Templeton wrote, his experiments had showed that bottlenecks could occasionally produce very strong sexual isolation.

Other support for the founder–flush model came serendipitously. In 1992, a team of marine biologists at the Woods Hole Oceanographic Institution in Massachusetts realized that a case of bottleneck speciation might have been going on under their very noses. They had *Nereis* worms in their lab for monitoring the effects of toxic substances. The worms had come a long way. In 1964, five or six of the worms had been collected on a beach near San Francisco and grown into a colony of thousands in a laboratory in California. On 5 December 1986, four pairs of these worms were taken to the east coast, to be the founder stock of a large colony at the Woods Hole institute. Over some 25 years, therefore, the *Nereis* had gone through two consecutive founder–flush periods, with extreme bottlenecks of only a few individuals.

When James Weinberg of Woods Hole realized that the worms in his aquaria were ideal candidates for founder–flush speciation, he set up an experiment. First, he went back to the Californian coast to collect more specimens at the site where they were originally found, back in 1964. Then he tried to mate these with his lab worms. The results were 'remarkable', Weinberg said, when he and his team reported them in the journal *Evolution* in 1992. First of all, the worms had literally drifted apart to the point where they had irreconcilable differences. When a male and a female were put together, they viciously attacked each other with their long, serrated jaws, and only now and then grudgingly mated. But those matings were far from successful. The males, which normally look after the brood (who would have thought that of the humble polychaete worm?), abandoned them, ate them, or generally were lousy fathers. Embryos rotted or were eaten by other kinds of worms. In the end, none of the mixed marriages produced any children.

It was an amazing case of a new species forming in less than a hundred worm generations, apparently under a regime similar to Carson's founder–flush model. Although the authors of the

Evolution paper tried to retain a sobre, business-like scientific style, it is clear that they were very excited when they wrote it. 'Very rapidly,' 'strong evidence', and 'very significant' are some of the terms in their conclusion.

Cracks in the bottles

By the late 1980s, bottleneck speciation was as firm a theory as it would ever be. Clever theorists had confirmed Mayr's and Carson's intuitions and proved that, on paper, it could work. Niles Eldredge, Stephen Jay Gould, and their many followers had crowned it as the speciation model that was best compatible with the fossil record, and then there were the experimental data: Templeton's work with *Drosophila mercatorum*, several other banana-fly experiments that had given mixed, but at least sometimes positive results, and, of course, that stunning thing with those Woods Hole worms. Biogeographers, meanwhile, continued adding data that were fully consistent with the model. Like Mayr, naturalists were finding time and time again that deviant populations or daughter species were often found in small pockets at the edge of a species' distribution range.

But there were bottleneck dissenters. Nick Barton, a population geneticist at the University of Edinburgh, had always been dead against it. One reason, he wrote, was because a bottleneck would get rid of almost all genetic variation. In the ancestral population, there might be a dozen alleles at each gene. The bottlenecked population would inherit a very poor sample of that variability—it would be lucky to get away with only one or two alleles for every gene. The fewer alleles, the less raw material there would be for evolution to work on. To make matters worse, Barton pointed out that it would not be the perfect breeding ground for new variants either. As Darwin had already remarked in the *Origin*, 'fewness of individuals will greatly retard the production of new species . . . by decreasing the chance of the appearance of favourable variations'.

So, Barton argued, how could such an impoverished population possibly change into a new species? There simply would not be sufficient critical mass for evolution to work. More likely, it would die out due to inbreeding or remain a genetically hollowed-out skeleton of its large parent-population.

At the same time, evolutionary geneticists began taking genetic fingerprints from species that supposedly had evolved in the bottleneck fashion, such as Hawaiian banana flies, the Galápagos finches, and a few other candidates. Given that these species should have started out as tiny populations, they should not have been able to carry with them much genetic diversity from their ancestral population. So they expected to find only one or two alleles at each gene. They had no such luck. In the Galápagos, for example, a team of American, German, Japanese, and Canadian biologists took blood samples from 20 birds belonging to 6 of the 13 Darwin finches. They looked at one particular gene, prosaically called *Mhc* Class II *B* group 1. Given that each bird has two alleles for each gene, the absolute maximum number of different alleles in their sample would be 40, while the minimum obviously was one; they found 21. Because the rate at which this gene mutates is known, the researchers could calculate that most of these alleles had evolved more than 10 million years ago, long before the islands were colonized. For so many alleles to have survived the genetic bottleneck, the team concluded that there must have been well over 40 founder birds, probably a few hundred. So the first species of Darwin finch had not evolved due to a founder event. The same applied to the Hawaiian *Drosophila*. Ironically, it was Alan Templeton who discovered that most banana-fly species in the archipelago were just as genetically variable as the mainland species.

But what about that trump card of bottleneck speciation, the laboratory worms? Surely that pillar of the theory still stood? It did so until 1996, when James Weinberg, its discoverer, toppled it in a 'brief communication' in *Evolution*. Together with two geneticists from the University of California at Irvine, Weinberg had been doing protein electrophoresis (the same technique used on those Pyrenean cave beetles in Chapter 2) to check if any tiny genetic differences had already built up between the San Francisco coast population (P1 and P2), and the ones in the Woods Hole toxicology aquariums. They found no tiny differences. Instead, they discovered a huge genetic chasm separating the two populations. They were different for almost every protein the geneticists looked at. So different, in fact, that there was no escaping the conclusion that the two groups of worms, although they looked identical, derived from two completely unrelated species. According to the accepted rate by which mutations in these proteins accumulate, the two species must have gone their separate ways no less than 6 million years ago, rather than just 25. So when Weinberg had gone back to the Californian coast to collect the wild relatives of his lab worm, he must have accidentally picked up an unrelated species.

With the same sort of detachment that NASA Mission Control reserves for announcing the tragic loss of spacecraft, Weinberg drily stated that 'the population sampled in 1964 . . . belonged to a species different from P1 and P2. Our interest was raised by early evidence suggesting that speciation had occurred in *Nereis* as a consequence of founder events . . . The results herein reported do not support this conclusion'.

The anti-revolutionaries

In spite of these theoretical objections, and the fact that the worm story had turned out to be a red herring, many evolutionary biologists still felt that the bottleneck idea was a strong one. After all, there were the fly experiments that showed that *Drosophila* could be pushed well onto the road to speciation by genetically bottlenecking them, and there was Alan Templeton's *Drosophila mercatorum* work. A few other tests had also been done, all of which had met with at least some success.

But in 1993, evolutionary geneticist Bill Rice of the University of California at Santa Cruz had a critical look at all the evidence. Together with his student Ellen Hostert he wrote an influential review article for *Evolution* entitled 'Laboratory experiments on speciation: what have we learned in 40 years?' In this article, Rice and Hostert started off by dismissing the strongest evidence so far—Templeton's *mercatorum* experiments. The fact that Templeton had used parthenogenesis to produce his one-individual bottlenecks was too artificial to their taste. 'There appears to be little relevance of these experiments to the natural speciation process', they wrote.

With Weinberg's and then Templeton's work shunted to the sidings, the remaining literature proved not powerful enough to uphold the theory. In an experiment with *Drosophila simulans* published in 1985, J. M. Ringo and colleagues reported getting only an extremely weak trend for bottlenecked populations to decrease in their willingness to mate with the original stock. And two biologists named Lisa Meffert and Edwin Bryant did a small study on house flies, published in 1992, which gave some positive results but none survived statistical analysis. Not impressed by this literature, Rice and Hostert blandly decided that 'no bottleneck experiment has produced levels of isolation even approaching those needed for speciation'. Alan Templeton was not amused. Three years later (scientific publication is a maddeningly slow process) he published a polemical paper in *Evolution* in defence of bottleneck speciation and against Rice and Hostert's conclusions. 'The most relevant papers in the experimental literature are not cited by Rice and Hostert', and 'inconsistent criteria for evaluation are used', Templeton fulminated.

Templeton was annoyed to see his *mercatorum* data slain so ruthlessly. But other experts had earlier agreed that there were serious problems with his study. Because his parthenogenetic *mercatorum* females duplicated all chromosomes in their egg cells, their daughters had the same allele at each one of their genes. This genetic situation is known to cause all kinds of genetic aberrations, and because it does not usually happen in nature Rice and Hostert were not alone in viewing Templeton's results as quirky.

But in other ways, Templeton did have a point. He revealed that Rice and Hostert might have been just a little bit prejudiced in their discussion of bottleneck speciation. Some experiments by Jeffrey Powell, for example, had found beginnings of speciation. Two out of eight bottlenecked laboratory populations of *Drosophila pseudoobscura* preferred to mate among themselves, rather than with the flies in the large population from which they were originally derived. Rice and Hostert thought this 25% success rate was statistically insignificant, so they were not impressed. But Templeton was. 'Obviously,' he wrote, 'to dismiss all of this work as insignificant is inappropriate and represents a fundamental abuse of statistics.'

The experiment to end all experiments

It was not just Alan Templeton who was uneasy about Rice and Hostert's paper. Other, less-belligerent scientists also thought the two had overplayed their hand. One of them was Arne Mooers, a Canadian now based at the University of Amsterdam, who, like most evolutionary biologists, got involved in speciation research in a roundabout way.

When Rice and Hostert's provocative paper came out, Mooers was a graduate student at the University of Oxford, where he was teasing information out of evolutionary trees. From the way evolutionary family trees looked, he tried to tell how quickly new species formed and how quickly they went extinct again. 'We'd been looking at speciation and extinction', Mooers says. 'I'd been doing that for three or four years, and then I realized that I didn't know much about the mechanism of speciation. I'd been talking about speciation and extinction, different *rates* of speciation—without actually knowing what that meant.'

So when Mooers returned to Canada after his doctoral studies he teamed up with two evolutionary biologists at the University of British Columbia in Vancouver. One was Dolph Schluter, a speciation expert who had worked on Darwin's finches in the Galápagos Islands and was starting research into speciation in fish (which is the capacity in which we will come across him later in this book). The other was banana-fly geneticist Michael Whitlock. Together, the three of them started discussing speciation theories. 'It was actually the Bill Rice paper that got us thinking about founder effect [saying that] there were no good data. That was the strongest statement that they made in that paper. So we thought, we should follow that up; do we believe that?', Mooers recalls.

'Rice and Hostert did some strange statistics to dismiss those data', Mooers says. 'And that's when we felt that it had been dismissed maybe too quickly.' So Mooers, Whitlock, and graduate student Howard Rundle designed an experiment to end all experiments. They decided that it would have to be big, to have sufficient statistical power, and they also decided to go natural. For example, both Powell and Ringo had used so-called hybrid starting populations for their tests. That is, they had mixed fly populations from different places, to obtain a maximum genetic diversity. Theoretically, this should increase the chances of getting speciation under way. Mooers and his colleagues instead used a single stock of banana flies as their raw material, and they incorporated other measures in their experimental design to avoid forcing the issue—such as imposing just one bottleneck rather than a whole series of successive bottlenecks, as all the earlier experimenters had done. 'Our question was, does it still hold when we know we are not stacking the deck?', Mooers says.

It was going to be a massive enterprise. First Mooers selected a starting population. It was to be a laboratory population of *Drosophila melanogaster* that had always been maintained at a large size at a lab in Britain; and it was genetically variable, although it stemmed from a single locality in Benin, Africa. In

1995, Whitlock carried more than a thousand flies from this population to Vancouver.

The next year, Mooers recounts, they took a whole bunch of fly populations from the stock; 50 of them. 'We took pairs, virgin males and females.' This was to mimic the bottleneck—two founding individuals on each of 50 'islands'. 'Then we let them produce a new population. And that new population we let grow to about 400 individuals and then let them stay at 400 individuals for a year.' This second phase represented the 'flush', which was followed by a period of stable population size to give natural selection the chance to act. 'Every two weeks we would take them all and count out 400 individuals, 200 males, 200 females, put them back in, let them grow up for a couple of weeks', Mooers explains. So the population fluctuated around 400.

Eight months and some 14 fly generations later, the researchers thought the time had come to see whether any nascent species had formed. They did this by checking if a female from a bottleneck population preferred to mate with a male from her own population or rather with a male from the baseline population, which had not been bottlenecked. 'We just took a female and two males and see who mated with whom', Mooers says. This simple statement utterly obscures the amount of slave-work that needed to be done. Mooers, Rundle, and Whitlock did 100 of such choice tests for each of their 50 bottlenecked populations. So they had to push 1500 flies into little glass vials, observe 5000 courtships, and over 4000 copulations through a microscope, each of which lasted for up to one-and-a-half hours (don't try this at home).

In the end, Mooers' team came out of their lab empty-handed. None of the 50 experimental bottles had produced any hint of speciation. All the bottlenecked flies were just as ready to mate among themselves as they were to mate with the ancestral stock, from whom they had been genetically drifting apart for 14 generations.

In 1998, Mooers and his colleagues repeated their test in a year-long experiment with a new set of 34 bottlenecked lines. Again, the flies refused to speciate. Indeed, the researchers discovered that when males from the bottlenecked populations had to compete for females, they always lost out to males from the large population. This matched Nick Barton's prediction: bottlenecked populations were unlikely to be an evolutionary success. 'The role of population bottlenecks in the speciation process should be de-emphasized', Rundle, Mooers, and Whitlock wrote in their 1998 *Evolution* paper. 'At least in *D. melanogaster*, single population bottlenecks do not often lead to the evolution of reproductive isolation. Other mechanisms of speciation should be given more credence.' 'I was personally disappointed that it didn't work', Arne Mooers admits. He points out that the bottleneck model should apply to almost all animals and plants, and hence it would be a general theory for speciation. 'A genetic idea that applied to everything in nature . . . great! I'm as drawn to those generalizations as the next person.'

Of course, the bottleneck debate does not end here. Ernst Mayr, for one, is unwilling to give up on his pet theory yet. He now says that *Drosophila melanogaster* may not be the most promising candidate for speciation in the bottleneck fashion. 'I would like to see these experiments performed with species that tend more to speciation and have more variation, but so far nobody has done that yet', he complains (in fact, Howard Rundle is now redoing the experiments with another *Drosophila* species). Even so, Mooers begs to differ with Mayr. 'We do know that you can select on almost anything in *melanogaster* and get quite a strong response. So there's a lot of variation . . . so I don't really buy that', he says. Also, the claim that because this banana fly has not spawned any new species in the wild it won't in a lab bottle either has been refuted since the early 1990s, when a sister species of *melanogaster* was discovered in Zimbabwe.

Although bottleneck speciation is not dead yet, it certainly has lost a lot of its appeal over the past decade. The difficulty of getting any laboratory speciation going using bottlenecks suggests that it is probably uncommon in nature. Peter Grant, an expert on Galápagos finches and a one-time bottlenecker, concluded in a 1998 book on island evolution that founder-effect speciation has now become the *enfant terrible* of speciation research. So let us take the current trend in evolutionary biology as an indication that bottlenecking is not a wise career move for a would-be species. The fact remains that geographical isolation is a key ingredient for speciation. How does it work? Why do isolated populations evolve an inability to mate with each other? Surprisingly, an obvious mechanism for allopatric speciation has been staring biologists in the face for ages, and yet it has only been appreciated since the 1970s.

CHAPTER FOUR

Seductive Theories

The Power of Sex

The male lyrebird of south-eastern Australia is as baroque as birds get. In the breeding season, it clears a circular mound of forest floor, and proceeds to put on its world-famous performance. It brings forward its two, long, lyre-shaped tail feathers and lowers them over its head, which is then completely obscured by the lacy fan of exceedingly delicate feathers in between. But the best part is yet to come: the concert.

Beautiful song is nothing special in birds, but the lyrebird is unique in its shameless use of samples. Interspersed among its own sounds are endless intermezzos of uncannily accurate imitations. It mimics the songs and calls of dozens of other bird species, and also any sound from its surroundings that takes its fancy. Barking dogs, for instance, or faraway sounds of engines, even chain saws and blows of hand axes are duplicated with hi-fi accuracy. And since natural history writers and photographers have started tracking the lyrebird, it has mindlessly added the mechanical clicks of camera shutters and laptop keyboards to its repertoire.

The male lyrebird is a good example of an excessively adorned male. Song, imitation, beautiful feathers, display, it has the lot. The female lyrebird, on the other hand, is drab and quiet. We have come across the same sexual inequality several times before in this book. In *Bombina bombina*, the fire-bellied toad that hybridizes with another species in a long belt across the European continent, only the males sing in deep booming voices, amplified by the air sacs in their throats. In birds of paradise, it is the males that have the splendidly coloured plumes that fascinated Ernst Mayr, Lord Rothschild, and millions of fashion-conscious women; and, again, it is the male cichlid fishes from Lake Nabugabo that are the brightly coloured sex. And what about those weird hammer-headed banana flies from Hawaii? Once more, the weirdness is restricted to the males.

Barring some exceptionally drab species, wherever you look in the animal kingdom, it is the same story. Females settle for the standard model, while males have all the accessories: spikes, crests, tufts, streamers, eyes on stalks, horns, grippers, and colourful attire. And it is they who sing, croak, roar, and dance. As we will soon see, this has implications for speciation. But first the pressing question; why all this exaggeration?

It was Charles Darwin who first provided an explanation. He had to, because the phenomenon was a severe obstacle to his theory of evolution by natural selection. The peacock's tail, his prime example, is an awkward and cumbersome appendage. It is larger than the rest of the body, which hinders the bird's flight and other movements. Moreover, its gaudy colours will attract predators much more easily than the subdued coloration of the peahen. So how could evolution have allowed this liability to develop? Surely, natural selection should have done away with it in no time, given its obvious negative effect on the bird's survival.

In 1871, a dozen years after *On the Origin of Species*, Darwin gave the answer when he put forward his second great evolutionary theory, in a book entitled *The Descent of Man and Selection in Relation to Sex*. Actually, the tome was two books rolled into one. The first and the last part were on human evolution, the subject that Darwin had felt was too controversial to put into the *Origin* (there, he hinted at it in a single remark, that 'light will be thrown on the origin of man and his history') but now apparently felt more comfortable with. The middle part of *The Descent*, which we are here concerned with, was only tangentially connected with the bits about human origins. It was on a type of selection that explained the peacock's tail, beetles' horns, and the angel suits of birds of paradise: selection in relation to sex, or sexual selection, for short.

Sexual selection, said Darwin, was the same as natural selection, but with the difference that it was not the physical environment that did the selecting but the opposite sex. No matter how cumbersome a male's ornaments, if they meant he would secure more females to mate with, he would get more offspring than less-exuberant males. As a result, his progeny would be over-represented in the following generation, and—inheriting the ornamentation—they, too, would be more successful. Sooner or later, all males in the population would carry the ornaments, the benefits in securing offspring far outweighing the trouble they may cause in day-to-day survival.

To a human-centred observer, this reasoning may not make immediate sense. If there is one male for every female, and each male teams up with a single female, there should be nothing to be gained from being a well-endowed male, because every male gets roughly the same number of children. But this sort of monogamy is almost non-existent, even in our own species. Polygamy, where males mate with more than one female (and/or vice versa), is much more common, and, indeed, it is in such species that sexual selection has left its trace most conspicuously (When Darwin, in preparation for his 1871 book, asked Mr Bartlett, the zookeeper in London, whether the male tragopan—a species of fowl—was polygamous, he was struck by the reply: 'I do not know, but should think so from his splendid colours'.) Even in so-called monogamous species, like many birds and mammals, including humans, extra-pair copulations—in other words, adultery—provide opportunities for sexual selection.

Darwin saw two ways in which the male's morphology could help to improve his mating success. The first was by battling with other males. An elk carries a heavy set of antlers on its head, which get entangled in undergrowth and cost him a lot of resources to grow and regrow each year. Nevertheless, it needs the antlers to combat and chase away other males. No antlers, no sex. Worse, with small antlers, no sex either. Only the males with the largest antlers, who win most of the fights, get access to the females and are able to secure some offspring.

Just as people can select the most aggressive fighting cocks, Darwin said, in nature the males that are best able to chase away rivals will be selected. 'The strongest and most vigorous males, or those provided with the best weapons, have prevailed under nature, and have led to the improvement of the natural breed or species. Through repeated deadly contests, a slight degree of variability . . . would suffice for the work of sexual selection', Darwin wrote. Now this was something that the world in the late nineteenth century could relate to. It was a pleasant extension to the 'struggle for life' that Darwin had already made popular as the stage against which natural selection was set. And the struggle between males was a very plausible mechanism for the evolution of male armaments. It explained why males are hoodlums.

But Darwin said more in the second part of his 1871 book. Because even though male combat would explain the evolution of horns, antlers, and other armour, it did not explain less-fearsome adornments like the tufts of pink fluff on the heads of redpolls, or the brilliant mauve testicles sported by male l'Hoest's monkeys, or, indeed, the peacock's tail. These males are not hoodlums. Instead, they are dandies. Darwin had an explanation for dandies too. It was the second way in which he thought sexual selection would work: by female choice. 'In the same manner as man can give beauty to his male poultry,' he wrote, 'so it appears that in a state of nature female birds, by having long selected the more attractive males, have added to their beauty.' Rather than physically battling each other for access to a female, these dandies used a different strategy: they would, as one author has written, woo rather than win her. And Darwin claimed: 'the females are most excited by . . . the more ornamented males, or those which are the best songsters, or play the best antics'.

Darwin knew his notion of females choosing dandies would not go down well in Victorian society. Foreseeing resistance, he wrote, 'no doubt this implies powers of discrimination and taste on the part of the female which will at first appear extremely improbable; but I hope . . . to shew that this is not the case'. He had no such luck. For a long time, female choice was considered a ludicrous idea. For a start, female animals were not considered capable of making such delicate choices between males, which all looked more or less alike—exactly the criticism that Darwin had anticipated. Moreover, everybody in those male-dominated days knew that it was the male who chose a female, and not the other way around. Females were considered universally passive organisms, regardless of their taxonomic affinity.

But there was another, fairer, criticism against female choice; namely why would a female choose a male with a long tail or bright blue balls? What good is beauty—if she had a concept of beauty at all—to a female? So for almost a century after Darwin had proposed it, female choice did not catch on. Rather, zoologists went to great lengths to explain dandies in the same terms as male combat. Perhaps by showing off their tails, bright colours, and strong voices, they said, these males would impress other males with their vigour and thus discourage them.

Only in the early 1970s, possibly aided by the centenary reprinting of Darwin's book, but also by women's lib, feminism, and the sexual revolution, did resistance wane against the notion of sexual selection by female choice. In that respect it is perhaps fitting that the first zoologists to start experimenting with female choice were the liberal-minded Scandinavians. In 1981 and 1982, Swedish zoologist Malte Andersson did some simple but ingenious experiments with the widowbird *Euplectes progne*. Females of this bird, which lives in the African savannah, are nondescript, mottled brown, sparrow-like birds. But the black males surpass even the peacock in tailiness, which in this species is almost ridiculous. Their thick tails are a whopping three times as long as the rest of their bodies. So when they fly slowly over the tall grass, displaying their exaggerated appendages by fanning them out underneath their bodies, they look more like a flock of UFOs than males defending their territory.

Andersson captured a few dozen of these territorial males on the Kinangop plateau in Kenya, and then proceeded to carry out some dextrous cutting and pasting. He clipped most of the tail feathers off some males, giving them unusually short tails of only 14 centimetres. With the leftover feathers and copious amounts of superglue, he then extended the tails of other males by some 25 centimetres. After this tail-styling, the males were released back into their original territories, as were some birds of which the tail length had not been altered. Andersson then surveyed the males' territories on a daily basis to see how the females would react to the treated males.

He noticed that the short-tailed and undamaged males attracted similar numbers of females. But the long-tailed males recruited twice as many females to nest in their territory. So, Andersson said, it certainly looked as if the females had a tail-fetish. They were preferring to have their eggs fertilized by males with super-long tails. But the alternative explanation remained that short-tailed males were intimidated by their long-tailed neighbours and gave up their territories. Not so, Anderson concluded. He observed that virtually all males retained their territories throughout the breeding season, regardless of tail length. So female choice did exist. Other Scandinavian ornithologists followed suit. The Danish zoologist Anders Pape Møller found that male barn swallows with artificially elongated tail streamers were preferred by the females. And his colleague Jakob Höglund painted white correction-fluid on the tail feathers of the male great snipe to make them even whiter than they already were, and was delighted to notice that this gave them better chances of seducing females.

Nowadays, female choice is a well-established fact. Numerous studies have shown that, rather than being passive bystanders, females often are fastidious, carefully selecting their partners on the basis of small differences in their looks and behaviour. But debate still rages over the reason why females do this. What do they gain when they pick a showy male? Do long tail feathers, red crests, and melodious song actually say something about the males' other qualities, or are these just irrational female whims?

Good taste or good sense?

The current controversy between the alternative causes behind female choice is nicely summed up in this quote from Darwin:

> The females are most excited by, or prefer pairing with, the most ornamented males, or those which are the best songsters, or play the best antics; but it is obviously probable, as has been actually observed in some cases, that they would at the same time prefer the most vigorous or lively males.

So do females gauge male 'vigour' (that is, good-quality genes) or is it enough for them just to be 'excited by' arbitrary sexiness? The first school of thought is referred to by cognoscenti as the 'good-geners' while the second camp is known as the 'Fisherians', after their patron saint, R. A. Fisher. We have met Ronald Fisher before (Chapter 3). He was one of the mathematical biologists who put evolution into the straightjacket of rigorous algebraic formulas, and laid the foundation for the Modern Synthesis. In his most famous book, the 1930 classic *The Genetical Theory of Natural Selection*, he tried his hand at sexual selection as well.

Male sexual ornaments could evolve quickly and unstoppably, simply by females acting on impulse, said Fisher. This is how he reasoned. Suppose some females in a hypothetical species have a genetic tendency to be more attracted to males with long tails. Now if at a certain point a mutant long-tailed male appeared, he would be able to father more offspring than the plain males to which all females were indifferent. The children of the choosy females who had mated with their long-tailed freak would on average inherit both the long-tail genes and the genes for long-tail preference.

Because the long-tailed sons would again have a benefit, the next generations would see an increase in long-tails among the males and an increase in long-tail preference among the females. Soon, every male in the population would have long tails and all the females would like them. If, at a certain point, a mutant male would be born with an even-longer-tail, the same process would repeat itself. Fisher wrote: 'The two characteristics affected by such a process, namely plumage development in the male and sexual preference in the female, must thus advance together, and . . . will advance with ever-increasing speed.' Fisher's scenario has since become known as 'runaway' sexual selection. Several Fisherians have performed fancy computer simulations and algebraic feats of skill that showed that their master was right. It could indeed work that way—at least until the male's ornament became so cumbersome that his sexual advantages were counterbalanced by his reduced survival.

But the good-geners are not impressed by that. To them the Fisherian theory is just a cerebral exercise that works on paper, but which they find gratuitous. They rightly say: 'Wouldn't it be much more plausible that there is some perfectly good reason for females to choose handsome males? Surely good looks translate directly to good genes?'

By now, there is some evidence that good-geners have a point, and that females are not always mindlessly following their whims. One particular thing that stands out is that females may use their suitors' showiness to measure their resistance against parasites. Anders Møller, for example, who earlier had shown that female barn swallows actually do select males with long tail streamers, in 1990 discovered that these long-tailed males were also the ones with the least blood-sucking mites on them. Apparently, healthy, mite-resistant males had more resources left to grow their feathers long and beautiful than scrawny ones. And their resistance against mite infestation was inherited by their offspring, even if these were raised by foster parents. So for a female barn swallow at least, choosing a long-tailed male is not a Fisherian fancy. She goes for his good genes.

Mark Boyce, a zoologist from the University of Wyoming, did a similar study in sage grouse. During the breeding season, in the plains of Wyoming and surrounding states, these birds convene at dawn on so-called leks. Derived from the Swedish word for 'play' (a linguistic legacy of the Scandinavians), lekking is serious business. A few dozen males will display all their charms for the passing females. They raise their spiky tail feathers into a vertical halo. They accentuate their black bibs by blowing up their white ruffs to reveal inflatable air sacs of bare yellow skin on their throats. To top it all off, on their necks and heads they erect bunches of black wiry plumes. Thus decorated, they proceed to wave their tails, throw back their heads, and prance about while producing bubbling sounds by inflating and deflating their air sacs.

Females mingle with these hyperventilating clowns and inspect the goods on offer. Once a female has made her choice, she squats in front of the male and permits him to inseminate her. Some males are very lucky and are allowed to mate with up to 30 females in a single morning, whereas other males are hardly visited at all. What Boyce and his colleagues discovered is that the males that were granted the least sexual favours were the lousiest ones—literally. They were most plagued by bird-lice. In fact, the lice were clearly visible on the birds' yellow air sacs. Were these spotted by the females? To test this, Boyce painted fake dots on the air sacs of healthy birds. Sure enough, their sex appeal plummeted. Good genes again.

But the Fisherians can also claim some experimental support. When Nancy Burley, of the University of Illinois in Urbana, was studying the behaviour of Australian zebra finches, she noticed something peculiar. She had given all the males in her experimental aviaries coloured leg-rings to be able to tell them apart. But she soon found out that she was not the only one who used these rings to discriminate among the male birds. The female zebra finches did, too. In fact, they preferred to court males with bright red rings around their legs, whereas they seriously disliked any males sporting green or blue rings. Stimulated by this serendipitous result, Burley (who is now at the University of California in Irvine) has since been experimenting with other artificial adornments on her finches. When she stuck little paper hats on their heads, she again found that females went for males with red headgear, and ignored the ones with green attire. In a latest experiment, published in 1998, Burley put tall white artificial crests on the birds' heads, a type of adornment that occurs in many other groups of birds—but not in zebra finches or any of their relatives. Again, crested males were considered irresistible.

In nature, needless to say, female zebra finches do not normally encounter males in paper hats, or crested ones, or ones with plastic rings around their legs. So these females apparently have a preference for male ornaments that do not really exist. Whatever their reason for falling for these dolled-up males, they certainly are not selecting any good genes. It must be purely aesthetic.

A male estrildid finch with a fake crest on its head, which makes it more attractive to females. Redrawn from a photograph courtesy of Nancy Burley, University of California at Irvine.

As usual in biology, when experiments support alternative hypotheses, the hypotheses may not be mutually exclusive after all. Could it be that Fisherian and good-gene sexual selection work at the same time? Could one type kick-start a bout of sexual selection and the other finish it off? Maybe male ornaments in some species are more likely to have evolved through Fisherian evolution, while other species are more prone to the effect of good genes? It is time for a reconciliation.

Speciation by fashion

Recently, such reconciliatory notes have sounded from University College London, where theoretician Andrew Pomiankowski has been doing a lot of thinking, modelling, and computer crunching. In a series of papers published between 1993 and 1995, together with Møller and the Japanese theoretical biologist Yoh Iwasa of Kyushu University, he developed an integrated view on what sexual selection can do to populations.

Initially, these authors wrote, a mutant male needs to hit what they call a female's 'sensory bias', as did the leg rings on Nancy Burley's zebra finches. For example, we humans have only three kinds of colour-detecting cell in our retinas, which is why a television screen is made up of only red, green, and yellow dots. If birds had TVs, they would need up to seven different colours to give realistic images, because birds have that many different types of detectors in their eyes. They live in a dazzlingly colourful world, which is why a mutant male bird with an unexpected bright colour (red, say) is likely to provoke a strong response in at least some females.

Once the red-colour gene and the female preference for it starts spreading, the stage is set for Fisherian exaggeration: a gene for 'even-redder' will spread, quickly followed by a gene for 'very very red'. At some point, a glass ceiling takes effect. The males cannot get any redder any more, because this would attract so many predators that it would outweigh the sexual benefits of redness. At the same time, when all the males in the population have maximum redness, the female's genetic predisposition for choosing the reddest males will start to lose significance. After all, there is no more cream to be skimmed, because all the males are equally attractive. The females evolve an indifference to red.

At that point, one of two things might occur. Once the female preference is reset to its original naivety, wrote Iwasa and Pomiankowski in a 1995 paper in *Nature*, the scene might be set for another bout of Fisherian snowballing, but this time in the opposite direction: if a mutant green male appears, it may set off rapid evolution into greenness (again, because a green male hits another sensory bias in the females). And so on, and so on, into cyclical eternity.

But another possibility is that at this point good genes will kick in. For example, many chemicals that animals use for pigmentation—such as carotenoids for red colour and melanin for black—also play a crucial part in animals' immune systems. A bird infected by parasites cannot afford to spend all his red pigment on tinting his feathers. As a result, it will be somewhat paler. So female birds will be forced to continue choosing the reddest among their suitors, and this will prevent Fisherian evolution moving in a different direction.

Exactly which of these two paths is followed depends on the direction of sensory bias, but also very much on the species' promiscuity. In relatively monogamous species, a male does not gain all that much advantage from being sexy. It may breed a bit earlier than another male, and it may score one or two adulterous copulations, but an attractive male will be able to exploit its charms much less than a male in, for example, a lekking species. In such polygamous species, it is not unusual for the sexual favours of the majority of the females to go to a single Casanova. As a result, the sort of cyclical, rapid, Fisherian evolution that Iwasa and Pomiankowski simulated on their computer screens is most likely to occur in more polygamous species, where runaway selection will proceed much faster and there will not always be enough time for good genes to butt in.

As Iwasa and Pomiankowski write in their *Nature* paper, 'cyclic evolution . . . will cause allopatric populations to fall out of phase quickly and evolve distinct sexual phenotypes'. And this is where we get back to the core subject of this book: the evolution of new species. In other words, these researchers are saying that Fisherian evolution is so unpredictable that, in allopatry, it is very unlikely to proceed in parallel. Isolated populations will inevitably diverge in their sexual ornaments *and* their sexual preferences. The implication is clear. geographical isolation can easily produce animals that do not recognize each other as sex partners any more. Fisherian sexual selection should produce new species, and a journey through the natural history literature supports this idea.

Wondrous willies

Plates 68 and 69—the sunbirds—of *A Field Guide to the Birds of West-Malaysia and Singapore* by Allen Jeyarajasingam are a dazzling confusion of colour. The superb paintings by artist Alan Pearson capture the kaleidoscopic plumages of these oriental birds as closely as is possible in two-dimensional water colours. The male black-throated sunbird has 'a maroon mantle, iridescent violet crown, rump and tail'. The female is olive grey. The male of the closely related crimson sunbird has 'bright crimson head and mantle, iridescent green forehead, iridescent violet tail'. The female, again, is olive grey. From the opposite page proudly stares the male of a third species, the red-throated sunbird, which has a light brick-red throat, dark maroon-red sides of the head, green crown and neck, deep ultramarine tail, and a yellow breast. Once again, its mate is largely olive grey. It's the same for all 11 sunbirds that occur in this small region in South-East Asia. The females are virtually indistinguishable, drab and greenish, while the males have turned themselves into an abstract art exhibition. Even the male of the so-called plain sunbird has a bright green spot on its forehead.

The pattern is familiar. Like sunbirds, many other groups of animals have species that are distinguishable mainly by their 'secondary' sexual characters (those not directly involved in the act of mating). The sunbirds' American counterparts, the hummingbirds, are an example. All species of pheasants, ducks, and finches, also differ from one another mainly by the feather ornaments of the males.

Lest you think these pages are getting very bird-dominated, the same applies equally well to other animals and to male exuberance other than feathers. Male fireflies (not really flies, but beetles) have species-specific patterns in the amorous messages they flash in Morse code. Male toads, frogs, crickets, cicadas, and grasshoppers all have songs that allow biologists to distinguish between otherwise very similar species. Courtship rituals are equally diverse. A male of the blenny-fish *Tripterygion*, for example, performs 'figure-eight swimming' in front of the female while she is spawning: *T. tripteronotus* does small eights, *T. xanthosoma* does extended eights, *T. melanurus* is specialized in slanted eights, while the male of *T. delaisi* does not get anywhere with his mate unless he performs a hotchpotch of all the figure-eights he can think of.

Even many 'primary' sexual characteristics appear to be subject to the same laws. Bill Eberhard, a zoologist at the Smithsonian Institution's Tropical Research Institute in Costa Rica, has demonstrated that the penis is not just a syringe for depositing sperm into females. Instead, he claimed in his 1985 book *Sexual Selection and Animal Genitalia*, it is an 'internal courtship device'. A detailed drawing of the male genitalia of the chicken flea, for example, is a profusion of plates, hairs, combs, spikes, bulbs, and spiralled wires, bearing an uncanny resemblance to an exploded view of some delicate piece of nanotechnology. It's a marvel of organic engineering, commented Eberhard, adding that 'it is just too fantastic to believe that such complicated machinery is necessary to perform a mechanically simple function'.

Chicken fleas are not unusual. Everywhere in the animal kingdom phalluses resemble more a diverse array of extraterrestrial kitchen appliances than they do a simple injecting organ. After scrutinizing all the hypotheses that had been offered in the past as explanations for all this penile decoration, Eberhard had to conclude that the anatomical decadence of penises evolved just like bird plumages, frog calls, and firefly flashes: through Fisherian evolution by female choice. The stimulation, in this case, was not visual or auditive, but tactile. This suggests that the contact between male and female genitalia during mating is better viewed as an extended courtship than as the act of fertilization. When the female does not like what she is feeling, she may decide not to use his sperm, cut short the copulation, and give her mate the slip.

Just like male bird plumages, penises differ tremendously between related species. Many taxonomists routinely use genitalia to identify species that are otherwise very hard to tell apart. Nematode worms, for example, are notoriously difficult to distinguish. As one nematologist has remarked, 'they are all narrow at one end, a bit thicker in the middle, and narrow again at the other end'. But their penises, amply making up for their unimaginative body shapes, are sometimes larger than the rest of the body. They are sculpted, rigid, asymmetrical claw-like things that even Darwin had trouble believing were caused by sexual selection. And in each nematode species, the penis is shaped differently; the female genitalia, though, are all similar. Eberhard listed dozens of examples in his book where the male genitalia have undergone as extreme an evolution as have sunbirds' plumages. Hoofed animals have penises that range from short stumps (deer) through long thin things with spikes (antelope) to the corkscrew-penis of the mouse-deer. Many other mammals, insects, snails, some fish, snakes, earthworms, mites, harvestmen, millipedes, woodlice, pseudoscorpions, spiders . . . 'no single major group in which internal fertilization is common fails to show this pattern', he wrote.

So not only courtship looks, sounds, and behaviour but also the shapes of penises have evolved independently and very rapidly in different species. This conclusion by itself does not say much. It shows only that many species differ in their mating signals and even in their genitalia, and that these things have probably passed through the Fisherian beauty parlour. It does not prove that sexual selection was the driving force behind their speciation—all this evolution could have happened after speciation took place. But the evidence, though still circumstantial, suggests it happened before. For instance, the newly discovered Zimbabwean banana fly and its sister species *Drosophila melanogaster*, which we came across in the context of bottleneck speciation in Chapter 3, have very different courtships. Banana flies produce what drosophilists euphemistically call a 'song': the males vibrate their wings, producing a series of buzzing sounds. Because the two species have very different songs, females of the one species will never mate with males of the other, and vice versa. But when Mike Ritchie of St Andrews University in Scotland forced the two species to interbreed, he found that, once their initial repugnance was overcome, they produced hybrids that were fully viable and fertile.

Ducks are another case in point. Female ducks are mostly brown, whereas drakes often have exuberant colours, crests on their heads, and exaggerated tails. Female mallards and pintails, for example, look very similar, whereas their drakes are as different as day and night. A male mallard has a yellow bill, a green head, a white collar, a chestnut breast, and two tiny curled-up black feathers at his tail; a male pintail has a grey bill, brown head, white breast, and tail feathers that have been extended into a spike 12 centimetres long. The two species, which are the most common ducks in the Northern Hemisphere, can be crossed in the laboratory without difficulty. The hybrids are completely viable and fertile. Yet, probably due to their different sexual signals, hybridization in nature is extremely rare.

Cases like these seem to prove that sexual signals keep species apart and hence cause speciation. But some still think the evidence for this is very flimsy. As Roger Butlin of the University of Leeds in England points out, cases like the Zimbabwean banana fly and the mallard–pintail might be rather special. According to this expert on animal hybridization, lots of species that are supposedly separate due to their sexually selected looks actually do engage in excessive intercrossing. So how could these devices ever be crucial in keeping species separate? Among birds, he says, it is in those polygamous hummingbirds, grouse, ducks, and birds of paradise that hybridization is especially rampant.

Polygamy breeds species

So we're back to birds. This is where we meet Anders Møller again, the Danish barn swallow experimentalist, who now works at the Pierre and Marie Curie University in Paris. In 1998, Møller and his Spanish colleague José Javier Cuervo of the Doñana Biological Station in Seville published results of their analysis to find support among birds for the idea that sexual selection drives speciation. They reasoned as follows. If arbitrary Fisherian sexual selection is sometimes responsible for speciation, one would expect it to be especially effective in polygamous birds. Delving into more than a hundred ornithology field guides and visiting some of the largest bird collections in Europe, they came up with a long list of species. For each, they recorded whether the males are ornamented, and whether the species is monogamous or polygamous. This information they scribbled into an evolutionary tree for all birds that two American ornithologists had set up in 1990 on the basis of DNA differences.

Møller and Cuervo then pored over the annotated tree to identify branches where ornamentation apparently had evolved, and found 70 such events. Next, they counted the numbers of species in the branches of the tree above the place where ornamentation had evolved. Most of the time, they discovered, the evolution of male decoration was followed by a flurry of speciation. On average, ornamented groups of birds had almost twice as many species as plain birds. And they also noted that these ornamented birds were mostly polygamous. It is a tantalizing result, which gives the impression that, at least in polygamous birds, sexual selection has indeed been the driving force in their speciation. Says Møller: 'That appears to be the case, and my interpretation of this is that you have perhaps a more intense directional selection and therefore more rapid divergence.'

But remember that Roger Butlin asserted that these highly polygamous birds are also the ones in which species barriers are the weakest; that is, where the most hybrids are formed. And a lot of hybridization could imply that two emerging species might easily mingle again when they come in contact. Møller thinks this is not a serious problem. First of all, he says, 'It's very, very important to distinguish between hybridization in nature and in captivity. Because in captivity, for example, you can interbreed [different species of] finches without any problems and the same applies to all the pheasants and all the ducks'. Still, he admits that he has also spotted the tendency for polygamous birds to hybridize a lot. In these species there is strong competition between males. As a result, Møller explains, males are probably more inclined to be sex offenders. The female might discriminate against mates with the wrong sexual signals, but the males will not always take no for an answer. 'Perhaps there is even stronger selection on the males to copulate with anything that moves', he says.

In any case, the analysis by Møller and Cuervo seems to indicate that this hybridization problem does not really negate the tendency for polygamous bird species to proliferate in numbers. In fact, Møller's analysis is only one in a series of studies of birds which point in the same direction. If the results hold up, sexual selection would appear to play a more important part in the origin of species than has been appreciated. And recently some of sexual selection's more sinister workings have been revealed, which may have an even stronger bearing on speciation.

The dark side of sex

The image of Fisherian sexual selection is a pleasantly cooperative one. Males show off, females shop around, and choose the ones they fancy. There's laughter and tears but nobody gets hurt. This is because males give females what they want, and vice versa. But in the last section we did meet one instance of males overstepping the line— if Anders Møller's suspicion is correct that polygamy also causes male birds to become rapists. Apparently what is good for the male is not always good for the female. In such cases, there is no longer cooperation: there is conflict.

The more intimate the contact is between a male and a female, the more opportunity there is for battling out any sexual conflicts. While the female keeps her distance, selecting which males to mate with, she is safe. But once she has allowed a male to insert his penis inside her, she is open to sexual manipulation. What sort of manipulation? Well, an enforced chastity of sorts. As has been said before, wantonness is widespread. Males mate with more than one female, but the reverse is also true. Females mate and re-mate. Even the proverbial mayflies, which may live less than one and a half hours, manage to get inseminated by more than one male. For females, promiscuity pays. Although a single suitor can fertilize all her eggs with sperm to spare, multiple matings might give her an additional way of selecting good genes, by allowing the sperm in her body to compete for vigour, swiftness, and navigational qualities. It may also guarantee a maximal genetic variability among her offspring, which will increase the chances of at least some young surviving.

To a male, a female's polygamous intentions are not good news. His interests would be best served if she used only his sperm and did not re-mate. So males have evolved an entire box of tricks to keep females on a sexual leash. Some males simply make the coitus last as long as it takes for the female to lay her eggs, which in some insects may be for over two months. Another common dodge is for males to leave a so-called 'mating plug' in the female's vagina after mating, which may prevent her from copulating with other males. This is a common practice in rodents, snakes, and lots of insects.

Drastic methods are used by the males in bedbugs, who resort to artificial insemination in a desperate attempt to guarantee that females use their sperm. With his penis, which is shaped like a hypodermic needle, the male bedbug pierces the female's body and then, with surgical precision, injects his sperm right where he wants it: near her ovaries. Adding insult to injury, the sperm then swim right through the female's body, penetrating tissue, even wriggling their way through individual cells, to find the female's eggs, which they then fertilize.

More sophisticated manipulation happens chemically. Besides sperm cells male ejaculates contain hundreds of proteins, sugars, and other substances, many of which serve to wage a chemical war against the female's adulterous inclinations. Once again, the best knowledge comes from the banana fly. *Drosophila melanogaster* males supplement their sperm with a cocktail of antiaphrodisiacs. A hormone called 'sex peptide' triggers a change in the female's sex drive. For up to seven hours after mating, she becomes uninterested in mating. The enzyme esterase-6 does the same, but it has an effect that lingers on for much longer. And at least two more, so far unidentified, compounds have similar effects.

Another means for males to get their way is to use the inside of the female to target other males' sperm directly. Besides the hormones that turn their mates frigid, banana fly males have components in their semen that kill old sperm from previous males. After a day or two, sperm cells inside the female become vulnerable to these substances. Why only after two days? Well, otherwise a male might kill its own sperm as well (which, after two days, should have done their job).

To the female, all this chemical harassment is unpleasant. Not only does it interfere with her sex life—doped-up, she cannot be the sort of swinger she may want to be, or, rather, exert the choice she might—but the actual chemicals are often toxic as well. Semen of the housefly, *Musca domestica*, leaches a hole in the female's vagina, which allows the antiaphrodisiacs to pass into her bloodstream. In the blowfly *Lucilia sericata*, the spikes on the male's penis are not just innocuous ticklers: they inject manipulative substances into the female. And just think of the traumatic copulations of bedbugs.

So sex is no picnic for females. In 1995, Tracey Chapman and her colleagues from University College London discovered just how damaging sex can be to the health of *Drosophila* females. They divided female banana flies into three groups. The ones in the first group were attended by genetically engineered males, which could produce only sperm and none of the additional fluids. Those in the second group were surrounded by another batch of transgenic males, but in these males the situation was reversed: they produced empty ejaculates, with no sperm, but only the fluids that normally go with it. The third group were control females, who had no males around. The results were clear: the females who were exposed only to the sperm and not to the noxious ejaculates lived just as long as celibate females—up to 35 days. Those who had to suffer the daily onslaught of chemical warfare in their bodies, however, lived no longer than 25 days.

Of course, during evolution, femaledom has not taken all this lying down. Females routinely shrug off males that copulate too long, they remove mating plugs, and in some species of bedbug the females have evolved an entirely new set of genitalia right in the spot where males prefer to pierce them. Similarly, females have probably responded by evolving countermeasures against the chemical assaults of semen. This sort of evolutionary conflict results in what biologists call a 'Red Queen' process. In Lewis Carroll's *Through the Looking Glass*, the Red Queen told Alice, 'Now, *here*, you see, it takes all the running *you* can do, to keep in the same place'. Appropriated for biology in 1973 by an American palaeontologist named Leigh Van Valen, the image of the Red Queen portrays what goes on when organisms evolve measures and countermeasures, just to stay—evolutionarily—in the same place.

It goes something like this. Female cuckoldry evolves. As a response, male claspers evolve to hold on to her after copulation. Females, not to be outdone, develop post-coital escapism. Males promptly evolve mating plugs. Females strike back by rejecting these plugs. The male then puts a substance in his semen to drug her into frigidity. The female retaliates by evolving an antidote. The male puts a second substance into his ejaculate, to back up the first. The female . . .

It is like a perpetual arms race, or, to use a friendlier metaphor, an evolutionary dance, where each move the man takes has to be matched by a countermove from the woman. That such dancing couples have often already covered a lot of dance floor can be gleaned from biochemical studies of the glands that produce the seminal fluids. We have seen that *Drosophila melanogaster* has no less than four different ejaculatory substances that prevent females from re-mating, and this number is likely to be vastly underestimated. When biochemists produce gels of the proteins that are produced by these glands, they see dozens of different bands, where each band may represent a male step and a female counterstep in their Darwinian tango.

The seed of speciation

It takes two to tango, but, as we shall soon see, to understand speciation it is necessary to explore the disastrous dance of two partners that cannot see, hear, or feel the other. What happens when male and female strategies evolve independently? This is what happens. When a Portuguese entomologist named A. Machado de Barros mated female tsetse flies (the insects that transmit sleeping sickness) from one part of Africa with males from another part of the continent, the females simply died. They had not had the opportunity to evolve in step, so the female was naive to the massive male onslaught of manipulative structures and molecules—and suffered the consequences.

A neat experiment to see this happening in the lab was done by Bill Rice, whom we met earlier (in Chapter 3) as one of the main critics of bottleneck speciation. In 1996, Rice published a paper in *Nature* describing an experiment he had done with (of course) *Drosophila melanogaster*. He had artificially prevented females from evolving countermeasures to the males' sexual trickery. How did he do this? Quite complicated; it involved a lot of fancy genetics. Basically, he raised a population in a laboratory cage in which male flies could experiment freely. They could evolve all the sperm-killing and female manipulation warfare they liked, without being hindered by female's countermeasures. This was because the females' genes were discarded each generation. In effect, as Rice wrote, the males were allowed to dance around a tethered female partner.

At the end of the experiment, which lasted for 30 fly generations, Rice ran his males through some tests and found that he had a bunch of lethal super-males on his hands. Unhindered by female moderation, they had evolved extremely nasty semen. It was 20% better at killing other males' sperm than normally. The females they mated with were not happy either: they died one-and-a-half times as fast as females mated with regular mates. Rice's experiments showed that in a mere 30 generations (just over one year of breeding flies), the inner environments of male and female reproductive systems can become quite incompatible.

By now it should be clear where this story is heading. Isolated, allopatric populations are just like those tsetse and banana flies that were not allowed to evolve in synchrony. Their dances will go in different directions, with the result that if a male from one population is crossed with a female from the other, they will no longer be compatible because they have been dancing to different tunes.

Molecular biologists have a further say in this. Since the early 1990s there has been a stream of publications showing that genes that function in sexual organs are the ones that evolve the fastest. In 1992, in a study led by Rama Singh of McMaster University in

Hamilton, Canada, it was discovered that the genes that produce the proteins in semen of banana flies evolve twice as fast as those in other organs. And in 1995, Singh proved that the same applies to genes in ovaries as well. In 1996, Miles Wilkinson of the University of Texas in Houston and his colleagues reported finding a so-called homeobox gene that did its work in the testes of rodents. Normally, such genes, which serve as 'master switches', turning on and off many other genes, evolve extremely slowly. But the one in mice and rats' testicles was frenetically ticking away its DNA clock as if towed along by the Red Queen.

The full significance of the sexual arms race and its bearing on speciation is only now beginning to dawn on biologists but the concept is already getting enthusiastic support. Roger Butlin, who claims he does not put much faith in regular sexual selection where speciation is concerned, is much more enthusiastic about the dark side of sex. 'I think that it might well be very important', he says. And Bill Rice writes: 'Conflict generates a . . . genetical chain reaction that may prove to be a major catalyst in the speciation process.' In an interview, he even called it 'the engine of speciation'.

It is not surprising that people are so enthusiastic. Not only is the field new and exciting but it also fills a gap that was conspicuously left open by Fisherian sexual selection. As we have seen, this latter type of sexual selection can explain why species differ in their plumage, call, or flash codes and hence will not mate. But it cannot explain why things go awry after mating has taken place, as happens when a donkey and a horse hybridize and produce the sterile mule. Another example are the banana flies *Drosophila simulans* and *D. mauritiana*. When they are crossed experimentally, they form hybrids, but the male hybrids cannot form proper sperm; they are sterile. Ernst Mayr thought it was just one of those things that accidentally drifted apart in allopatry.

But now, the sexual arms race is beginning to look like the culprit for such post-mating isolation. Actually, a 'three-way tug of war', as Bill Rice calls it, is a more apt description of the process, because males and females are not just fighting each other but males are also fighting other males in sperm competition. And the females, whose beleaguered reproductive system is staging all this hostility, have to come up with new countermeasures all the time. Just think of what those molecular studies are telling us. The chemical composition of semen, and probably also of sperm cells and eggs, are constantly undergoing rebuilding, to cope with the ever-changing attacks from other players in the three-way tug of war. It is not at all surprising that eggs and sperm that have gone through different tugs-of-war are no longer compatible.

Similarly, a hybrid will have genes deriving from different arms races. Genetically, it will be a mess: it will have some genes from the one species, other genes from the other species. So a hybrid male may have weaponry to kill other males' sperm, but it may lack the proper gear to prevent attacks on its own sperm. Or it may lack a genetic master switch that is supposed to turn on an entire suite of other genes. In fact, one important gene that sterilizes male hybrids of the banana flies *Drosophila simulans* and *D. mauritiana* has turned out to be this type of rapidly evolving, homeobox gene that functions in the testis.

The bottom line is that sex causes tremendously fast and dramatic evolution in animals. It changes the way animals look outside and inside; it changes the visual, auditory, tactile, and chemical messages they send out; and it seems fit to be a major propeller of speciation. After all, it produces the very hallmark of speciation (at least in the view of the modern synthesists): reproductive isolation.

Refurbishing geographical speciation

Now, more than half a century after Ernst Mayr returned from his Pacific voyages with a bag full of dead birds and a head full of viable ideas, speciation research seems to have come full circle. The skeleton of geographical speciation is still there, but where

Mayr saw founder events and genetic revolutions, we now see sexual arms races and Fisherian sexual selection.

There is a bit of irony in sexual selection's resurgence of power. Remember that one of its major champions, Ronald Fisher, was among the theoreticians on whose ideas the foundations for the Modern Synthesis were laid (see Chapter 3). Yet his ideas on sexual selection were never picked up by Mayr and Dobzhansky when they started thinking about how speciation worked. Even though Fisher's model was there for the taking, Mayr ignored it. When Mayr discovered that 'peripherally isolated' populations of birds—such as the spangled drongo and the paradise kingfisher (see Chapter 2)—had strikingly different plumages, he concluded that these were the accidental byproducts of the genetic revolutions that came with the foundation of these populations. The entire genetic constitution of the species was rearranged during and after the bottleneck and, said Mayr, some genes for sexual plumage patterns would just tag along. Of other sexual signalling, like penis shape, he also thought the differences were 'accidental'.

Later, in a 1992 retrospective, Mayr realized that the absence of sexual selection from the Modern Synthesis was 'surprising'. Anders Møller remarks: 'Many people knew what [Fisher] wrote, but whether it had any relationship with what was going on in nature was not so clear. In that time, sexual selection was considered a marginal subject.' Nowadays, sexual selection is no longer marginal; it has become fashionable mainstream evolutionary biology.

In the 110 years that have passed since Moritz Wagner proposed it, and Charles Darwin called it utter wretched rubbish, the theory of geographical speciation has come a long way. Isolation by itself does not produce new species, nor do genetic bottlenecks, it seems. Instead, two interacting forces do. Sexual selection and sexual arms races produce the reproductive isolation that Mayr felt was so important. And, as we shall see in later chapters, both are intimately intertwined with natural selection.

Even though the theory has changed so much over the years that Wagner and Darwin would not recognize it any more, geographical speciation is here to stay. Biology textbooks state and restate that it is the most widespread and universal mode of speciation. But even in the most geography-centred textbooks, there is always a small box that pays tribute to an alternative mode of speciation. In some plants, under certain circumstances, ready-made species are churned out while you wait. Harking back to the dark ages of evolutionary biology, the notion of instant speciation appears to be justified. And what's more, not just in plants either. This is what the next chapter will be about.

CHAPTER FIVE

Wham, Bam, Brand New Species

On the Instant Origin of Species

In a corner of the Botanical Gardens of the University of Amsterdam stands the bust of a bearded man. Around the pedestal, tall yellow flowers grow. The mossy statue portrays the Dutch botanist Hugo de Vries; the flowers at his feet are evening primroses, the plants that made him famous.

In 1886, while strolling in an estate near Hilversum, some 30 kilometres outside Amsterdam, de Vries stumbled on a fallow potato field, all overgrown with evening primroses. The plants were among the first specimens to have made their way into the Netherlands, after first having been introduced from their native Texas into England and Germany. But it was not his find of this recent addition to the Dutch flora that most excited de Vries. Among the plants that he saw in Hilversum were two very odd specimens. One had flowers with very short styles (the stalk on which the stigma sits). The other was a remarkable smooth-leaved form with 'much more beautiful foliage', and narrower, straight petals. De Vries, a 38-year-old plant physiologist who had just started to get interested in the then unsolved mystery of genetics, realized that here he had a good floral guinea-pig for him to start experimenting on. As he took some seeds from the evening primroses for his laboratory in the Botanical Gardens, the rest of his life began.

Over the years that followed, de Vries managed to breed seven more new varieties (rediscovering Mendel's laws of genetics in passing). In every batch that was sowed, a few strange individuals would spring up unexpectedly, very different in their size, the colour and shape of leaves and petals, the colour of stems and veins, and so on. He wrote: '[They] come perfectly true from seeds. They differ . . . in numerous characters, and are therefore to be considered as new elementary species.' Moreover, the new species was constant: if he fertilized the flowers of an aberrant plant with its own pollen, the offspring would be identical to the parent. De Vries thought he was on to something good. These were ready-made species, generated during a single generation in a creative genetic process for which he coined the term 'mutation'.

In 1901 and 1903, de Vries assembled all his work and his ideas into two large tomes entitled *Die Mutationstheorie*. In this major work he put forward his 'mutations' as a radical revision of Darwin's theory of evolution by natural selection. He saw no gradual change over many generations: it was wham, bam, brand new species. 'The new species originates suddenly,' he wrote, 'it is produced by the existing one without any visible preparation and without transition.' Darwin, echoing naturalists of old, had said that *Natura non facit saltum*—Nature makes no leaps. With a naughty wink to Darwin, de Vries and his followers adopted the name 'saltationists'.

Although de Vries is best remembered today as one of the rediscoverers of Mendel's laws of heredity, during the first twenty years after *Die Mutationstheorie* he was heralded mainly because he was the man who had done better than Darwin—by proving that evolution did not (or at least not only) proceed through natural selection. Most biologists in those days became convinced that the mutational leaps that de Vries had discovered were the true stuff of evolution. Even as recently as 1932 a prominent biologist wrote, 'All of Darwin's "particular views" have gone down wind: variation, survival of the fittest, natural selection, sexual selection, and all the rest. Darwin is very nearly, if not quite . . . outmoded.'

But soon the mutation theory began to show flaws. Other botanists discovered that evening primroses and their relatives are somewhat peculiar genetically. They often suffer from large-scale irregularities in their chromosomes, resulting in those freaky flowers. It seemed that what de Vries had called 'new species' were just deviant plants. It was not surprising, then, that de Vries and his saltationism were one of the first targets for the scientists of the Modern Synthesis. To them, saltationism represented the typical old-fashioned sort of biology that science needed to get rid of. The modern way was to think of evolution as taking place in populations. Natural selection or genetic drift would drive new alleles through a population. The mere appearance of a new mutant, however spectacular, was not yet speciation.

In Ernst Mayr's view, the saltationists made two mistakes: they ignored populations and they ignored geographical isolation. One modern proponent of saltationism was a particularly unfortunate recipient of Mayr's scorn. Richard Goldschmidt was a geneticist who, like de Vries, thought that the almost imperceptible natural variation was insufficient for the evolution of entirely new species. He felt that rare freaks (a fly with four wings, for example) might have a better chance. In his 1940 book *The Material Basis of Evolution* Goldschmidt coined the term 'hopeful monsters' for such mutants. Mayr agreed that such mutations occur now and then, but he did not think they would be of any relevance to speciation. 'They are such evident freaks that these monsters can be designated only as "hopeless"', he wrote. And in 1952, to expose that saltationism was naively ignoring what goes on in a population, Mayr went up to Goldschmidt and asked him how the population in which a new hopeful monster occurred would react to it. As Mayr recalled: 'He answered, after a considerable pause, "I have never thought of it that way".'

In Mayr's view, instant speciation was an impossibility, and it certainly could not happen sympatrically, right in the middle of an existing population. Any newborn monster would be a still-born monster and not hopeful at all. But there was one particular type of saltationism for which he made an exception. And that was a phenomenon called polyploidy. As the following section will reveal, this process churns out ready-made species all the time in plants, in a way that would have thrilled de Vries.

Double your genome today!

We will have to start off with another lesson in our genetics crash course. Remember that a plant (or animal, for that matter) usually has two copies of each chromosome: one from its father and one from its mother. Before sex cells are formed, each pair of chromosomes align, jumble their genes, and split up again, so that only a single copy of the recombined chromosomes is packed into sperm cells or eggs. So in pollen grains (which contain the plant's sperm cells) or ovules (which contain the plant's egg cells), the number of chromosomes is halved because there is only a single copy for each chromosome instead of a pair. But once a sperm cell fuses with an egg to start a new plant, the chromosomes are restored to their original number. Regular cells with two sets of chromosomes are called 'diploid' (from the Greek words *diploos* and *eidos*: 'double image').

Sometimes, this packaging business goes wrong, when the cells 'forget' to split the pairs of chromosomes before bundling them up into sex cells. The result is a sperm or egg cell with a double set of chromosomes. When this diploid sex cell fuses with a normal sex cell from another individual, the result is a baby plant with three sets of chromosomes. Unless they perform parthenogenesis, like the dandelions we met earlier in Chapter 1, such triploids are a dead end: the molecular machinery in their sex cells is not equipped to extract single sets of chromosomes from the triplets. Three is a crowd and triploids are sterile.

But what if a diploid, rogue ovule happened to be visited by an equally overendowed diploid pollen grain? This may be a rare chance event, but, if it happens, the result would be a seedling with four sets of chromosomes, a so-called tetraploid. And four sets of chromosomes *can* be evenly split down the middle to produce, once again, diploid sex cells. These can then start fertilizing other tetraploid plants in the vicinity or (if it is one of those plants that have both male and female functions within the same flower) each other. It's instant speciation. In a single generation, genome duplication produces plants that can no longer reproduce with their parents, but only among themselves.

The process is particularly efficient if it helps to hybridize two related species. Often, plant hybrids are sterile because the two sets of chromosomes are so different that they cannot properly line up. But in a tetraploid, each chromosome has one potential partner and forming sperm and eggs is no longer an obstacle. Moreover, hybrids are genetic intermediates, which may help them find their own niche without having to compete with the two parent species.

This is not science fiction. Biologists have known for more than half a century that polyploidy has been rampant in the plant kingdom. In 1950, George Ledyard Stebbins (the botanist of the Modern Synthesis and author of *Variation and Evolution in Plants*) estimated that a third of all plant species were polyploids, and in 1963 another botanist, Verne Grant, even claimed an estimate of up to 95% for some groups of plants. How did they arrive at such figures? Grant did so by drawing a simple graph. On the horizontal axis of his chart Grant put the numbers of chromosomes in pollen, while on the vertical axis he had the numbers of plant species that carried these numbers. The result was a line sloping to the right (many species with small numbers of chromosomes, and few species with very large numbers of chromosomes), and peaking at regular intervals. Grant noticed that all the peaks marked even numbers of chromosomes while all the valleys were at odd numbers—a tell-tale sign of a long history of recurrent polyploidy, because a doubling of any number, odd or even, gives an even number.

Many polyploid species with the same number of chromosomes may derive from a single ancestral polyploidization event, so the number of polyploid plant species is probably an overestimate of the total number of polyploid speciations. Nevertheless, Pamela Soltis, a botanist at Washington State University in Pullman, is convinced that polyploid speciation is very common in plants. She should know—she has seen it happen in her own backyard.

Soltis and her botanist husband Doug Soltis have been studying weeds called goat's-beards in the north-western United States. These graceful cousins of the dandelion are not native to that area. Three different species, known officially as *Tragopogon dubius*, *T. pratensis*, and *T. porrifolius*, were accidentally brought in from Europe in the beginning of the twentieth century. Since then, the plants, which thrive in abandoned fields, roadsides, and backyards, have flourished in and around towns. *Tragopogon pratensis* and *T. porrifolius* tend to remain restricted to the states of Washington and Idaho, but *T. dubius* has now spread over most of the northern United States. Although most people consider them nasty weeds, Soltis—understandably—thinks they're nice. 'They're beautiful when they're in flower but they're a pestilence when they go to fruit. So we tend to let them go in our garden and in our yard for as long as we can stand it and then we go out and pull them up.'

In 1949, one of Pamela Soltis's predecessors at Washington State University, Marion Ownbey, discovered 56 plants growing along a railway line near Pullman that looked decidedly odd. They were larger and seemed intermediate between *T. dubius* and *T. porrifolius*, which were also growing there. When he made microscopic preparations of the plants he discovered that they had 24 chromosomes, rather than the normal 12. They were polyploid hybrids, which, as Ownbey confirmed, could not be crossed with the parent species but only among themselves. In 1950, Ownbey classified them as a new species, *T. mirus*. In the same year, he also found polyploid hybrids between *T. dubius* and *T. pratensis*, which he named *T. miscellus*.

Although Ownbey stated that the censuses of his two new species were 'precariously small', the intervening half century has been good to the duo of neophytes. In 1990, Pamela Soltis found *T. mirus* alive and well in an area measuring 25 by 75 kilometres in Washington state. The other one, *T. miscellus*, was doing even better. It had by then spread over large parts of eastern Washington and western Idaho, where it was, after *T. dubius*, the second commonest goat's-beard. Around the town of Spokane, Soltis found it abundantly throughout an area no less than 790 square kilometres in size. In the town of Spokane itself grew an estimated 10 000 plants of this tetraploid.

Since then, Soltis has been keeping a close watch on the two new species. She has discovered that they spread not only by casting seeds around (like dandelions, goat's-beards have parachute seeds that sail in the wind for long distances), but also by multiple origins. 'We have actually, over the course of the last 10 years that we've been watching these things, seen a couple of new tetraploid populations pop up,' she says, 'so it seems to happen every time that they come into contact with each other.' DNA analysis is backing up this claim. Soltis has been studying the plants' chloroplasts, the green grains inside plant cells where photosynthesis takes place. Like mitochondria, chloroplasts are inherited through the egg only; a sperm cell does not contribute any chloroplasts to a baby plant. So by looking at the chloroplast DNA, Soltis was able to tell which species a hybrid tetraploid had had for its mother. Sure enough, some populations of tetraploids had chloroplasts only from one parent species whereas others had chloroplasts only from the other parent species.

So both new species have arisen at least twice, but Soltis thinks that a higher number may be closer to the truth. Using more refined DNA technology, she has now shown that most populations of the new species are close relatives of the parent plants that grow at the same spot. 'I guess I would give a minimum estimate maybe of around four origins of *T. miscellus*', she says.

'With *T. merus*, the other one, the number is higher. There are about five or six really good origins and then that could get up to maybe as high as 13 or so.' Another surprise was that the hybrid species appear to have secured their own niche. For example, on open slopes where *T. mirus* grows in the company of its parental species, *T. dubius* and *T. porrifolius*, the latter grows in wet and shady places, the other diploid grows in dry and sunny spots, while the tetraploid forms large bushes in all the intermediate spots, which are a little wet, or somewhat dry, not so sunny and not too shady.

It is a fascinating example of new species forming within a human lifetime. But their history begs one question. If all three species are from Europe, why did the new tetraploids not form there as well? Perhaps they did. Worse still, perhaps they were introduced with the original parents and simply not noticed until Marion Ownbey came along. Soltis, however, is certain that that cannot be the case. In Europe, she says, the different species do not grow together in mixed stands in such profusion as they do in North America. Europe is the best-studied region in the world, botanically. If tetraploids were there, her European colleagues should have discovered them long ago. Moreover, the tetraploids' DNA clearly shows that they are pure-bred, all-American goat's-beards. 'It's great', she says.

Polyploidy in plants was the only form of instant speciation that Mayr allowed. It was so obvious that he gladly accepted sympatric polyploid speciation as the alternative for his allopatric model. But it now seems that there are other instantaneous ways of getting new species. Recently, for example, a somewhat similar phenomenon has been revealed in animals.

Please keep right

A few years ago, large billboards along the German motorways encouraged sluggish drivers to stay in the slow lane. 'Please keep right!', it said, across a large photograph of a slowly slithering escargot. Whenever Edmund Gittenberger, curator of snails at the National Museum of Natural History in Leiden, the Netherlands, would pass one of these signs, he had to chuckle inwardly. Not because he would ever disregard the well-meant warnings of the German Federal Ministry of Transport, but because the snail, unlike its real-life brethren, was left-handed.

To mollusc curators, there is no greater crime than mistaking left for right. Just like gloves, shoes, and yoghurt, snails come in two forms: left-handed and right-handed. Right-handed snails, which are the type normally seen in nature, are coiled along an imaginary axis in a clockwise fashion. If the shell is held with the mouth facing the observer and the tip up, its mouth is on the right-hand side. In a left-handed, anticlockwise shell, the mouth is on the left. But, as Gittenberger observes to his chagrin, most people are oblivious to these subtler points of malacology. When a person draws a snail, he or she tends to draw it as a left-handed specimen. 'I once annoyed a painter by pointing out to her that the snails in her still-lifes were all coiled the wrong way', he says. And escargots are always of the right-handed variety, too, so the designer of the German billboard must have accidentally reversed the negative of the photograph. Or, worse still, he did so on purpose. A snail is a snail, he may have thought, and sacrificed scientific accuracy for compositional beauty.

But every now and then, nature itself makes the same mistake. Very rare left-handed escargots have been found. Most shell collectors would put such a freak straight into their collection, but in the early 1900s the German naturalist J. Meisenheimer kept a specimen alive for a slightly sadistic purpose.

Because snails are so asymmetric, they have to mate in a peculiar way. Their sex organs are stashed away behind a small hole on the right-hand side of their heads (if they are coiled the regular way). When two snails engage in copulation, they approach one another face-to-face and then rub their cheeks against each other's necks to bring the genital openings in contact. Only then, can mutual penis insertion take place (snails are hermaphrodites, which means they are male and female at the same time, although they normally do not fertilize themselves). So what Meisenheimer did was put his twisted escargot in a cage together with a right-handed specimen. With the two animals being each other's mirror image, they would not find their partner's genitalia where they expected them. So how did they solve this problem? To cut a long story short, they didn't. Living up to their proverbial phlegmatism, 'for days and weeks the animals fatigued each other in courtship, without achieving a final copulation', Meisenheimer wrote in 1912.

When Edmund Gittenberger came across this and similar references in the scientific literature, he realized that coiling gone awry might actually induce speciation in snails. If a mutation could cause a snail to be able to copulate only with other snails carrying the same mutation, the stage would be set for the evolution of a new, mirror-image species. Luckily for him, the genetics of snail coiling had already been known since the 1930s, when geneticists discovered that only a single gene is involved, with two alleles: one for right and one for left. Because the *R* allele is dominant (its effect overrides that of the other allele), *R/R* and *L/R* pairs of alleles code for right, while only *L/L* codes for left. But there is one catch: the gene does not have its effect on the animal itself, but on its eggs. So an *L/L* snail from an *R/L* mother will be right-handed, because its mother is right-handed. Mirror-image snails will appear only a generation later, as the offspring of the *L/L* snail.

To Gittenberger, this one-generation delay was a bonus, because it meant that the *L*-allele could penetrate a population without immediately causing left-handed snails. But at a certain point, suddenly many left-handed snails would appear out of nowhere, which could then mate among themselves and kick-start a new species. In 1988, he published his idea in a two-page note in the journal *Evolution*, entitled, 'Sympatric speciation in snails: a largely neglected model'. The last line of the paper read:

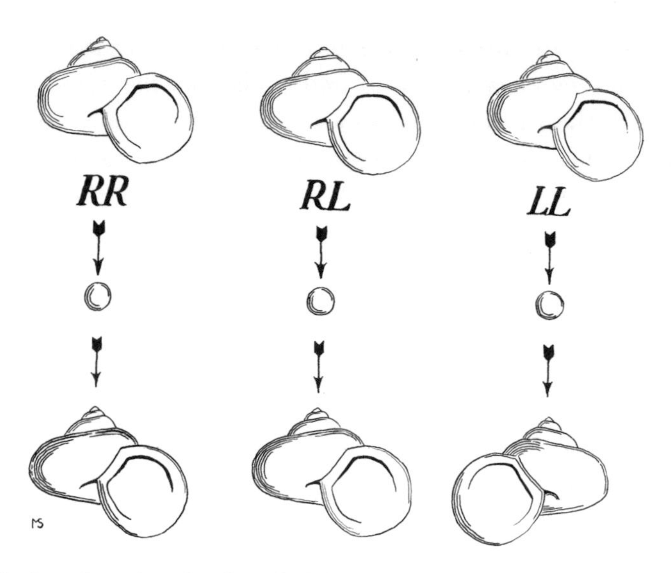

The heredity of snail-coiling. Left-handed snails are born from eggs laid by a snail that has inherited the *L*-allele from both parents.

'The population-genetic processes involved in this are complicated and deserve to be studied mathematically.'

In the years that followed, four biologists took up the gauntlet and did calculations to see if Gittenberger was right. In 1990, Michael Johnson of the University of Western Australia and two of his colleagues decided that he was not. '[It] is particularly unlikely to serve as the basis for sympatric speciation', they wrote in *Evolution*. Their calculations showed that the first few left-handed snails would be hard-pressed to find partners with the same coil. Because many may not find mates, the mutants would die out again as suddenly as they had appeared. A year later, however, Allen Orr of the University of Chicago thought otherwise in a third note on the subject in *Evolution*. His computer simulations had shown that now and then an entire population could be taken over by lefties. When this happens, Orr wrote, 'single-gene speciation is possible'. And when Gittenberger and his mathematically inclined colleague Eke van Batenburg did a large series of computer calculations themselves, they came up with the same conclusion. In small populations, left-handed snails could occasionally gain a foothold.

In his 1988 paper, Gittenberger had compiled evidence that nature, at least temporarily, lives up to the theory. Back in 1862, a colony of left-handed relatives of the escargot was found in a private garden in England. And in 1911, J. W. Taylor reported finding almost 2000 left-handed fossil shells of the so-called grove snail in geological deposits a few thousand years old. Apparently, these populations had not survived. Sometimes all members of a species are left-handed. For example, *Quantula striata*, a Singaporean snail famous for the flashing light it carries between its eye stalks, is right-handed but its close relatives, the *Dyakia* snails of Borneo, are lefties. A West African cousin of the well-known giant African snail is also left-handed, as are some *Ariophanta*'s from India and Indonesia. In fact, as one malacologist has remarked, 'there is a cornucopia of contrarily-coiled conchiform crawlies'.

Gittenberger points out, however, that it is crucial to distinguish between roundish snails like escargots, and slender snails. The latter type mate in a different way. 'The shell simply does not get in the way', he says. This means that slender snails can mount each other, move their heads close to each other, and copulate. Even a mixed left/right pair can mate, albeit with some difficulty, as several snail observers have determined. All they need to do is shift their heads a bit. The upshot is that instant speciation cannot take place in these slender snails, because a reversal in coil does not make them sexual outcasts. So instant speciation by coil will work only in the escargot-shaped snails. And left-handed species are extremely rare there. There are only a handful of examples world-wide, says Gittenberger. This means that snails that do the twist are as rare as slow-driving Germans.

Although coil reversal may occasionally trigger speciation sympatrically in snails and, on a regular basis, polyploidy does the same in plants, these things are usually considered exceptions to the allopatric rule. As mentioned before, most textbooks cite geography as the only wholesale species creator. Ernst Mayr is even more categorical. In 1963, he wrote: ‘The widespread occurrence of geographical speciation is no longer seriously questioned by anyone.’ And, today, he is no less convinced, when he says: ‘That is the exclusive one in mammals and in birds and in most animals and in most flowering plants.’

But some people are not so sure. Jeffrey Feder, an entomologist at the University of Notre Dame in Indiana, says: ‘Mayr has given the impression of: here’s the picture; this is what it is like; now you get some crayons and colour it in.’ Feder thinks there may be other pictures to colour in than just the allopatric one. He is one of the growing band of geography-dissenters who adhere to a heretical theory of speciation without geographical isolation. Such boundless speciation is what the remainder of this book deals with.

CHAPTER SIX

A Chronic Case of Rhagoletis

The Birth of an Evolutionary Heresy

The Natural History Museum in London is one of the most impressive museums in the world. Its huge crenellated building, occupying the sunny side of Cromwell Road, houses an estimated 90 million specimens. In the public exhibits, not much of the vast collection is visible. Most of it is hidden behind oaken doors marked 'STAFF ONLY' or 'NO ACCESS'. But the visitor who is fortunate enough to be allowed beyond those doors, leaving behind the noise of shouting dino-crazed schoolchildren and the hissing and puffing of pneumatic grasshoppers, enters the serene world of handwritten labels, formalin, and patient cataloguing.

And what a world it is. Imagine a maze where moth-eaten giraffes mingle with endless plumbing, and corridors flanked by thousands of wooden, glass-topped drawers containing millions and millions of pinned insects. Having been warned by the curator not to touch anything, the curious visitor might still be involuntarily drawn toward a cabinet bearing the enigmatic words 'Curculionidae—Palaearctic'. As she quietly pulls a drawer a fraction from its frame, batallions of tiny, metallic weevils appear. Each barely different from its neighbour and, yet, all different species.

Weevils are the Pinocchios among beetles. They have an extended snout, with two antennae attached somewhere halfway along it. As a group they are particularly species-rich. In fact, they may be the largest family of beetles, with over 60 000 species worldwide. Being a weevil-collector means that you have to be a bit of a botanist as well, because many weevils are extreme food specialists. They will feed on a particular type of plant, and nowhere else. Often you can find a certain species only by locating its food plant. Such species are termed 'monophagous' (from the Greek *monos* for 'only' and *phagein*, 'to eat').

A bearded weevil-specialist volunteers a guided tour. 'Weevils are the most fascinating animals on the planet', he says, with shameless 'curculio'-centredness. He opens a drawer with so-called 'apionids', tiny weevils, the shape of miniature shrews. 'This one, for instance,' he says, pointing at a 2-millimetre-long black one with yellow legs and antennae, 'is *Apion gracilipes*. It lives and breeds exclusively on zigzag clover.' Its neighbour, *Apion flavipes*, identical but for its dark antennae, has slightly different tastes: it lives on white clover and alsike clover. And there are hundreds of examples of such extreme food specialization in weevils. But weevils are nothing special as far as monophagy goes. Many other groups of insects also specialize in their eating habits. For example, of the 300-plus species of leaf-mining flies in Britain alone some 70% are monophagous. In the tropical rainforest, this situation reaches its pinnacle. The American ecologist Daniel Janzen has estimated that about half of all the butterflies and moths in the tropics eat only one species of plant.

In the late 1970s, an entomologist named Terry Erwin from the National Museum of Natural History in New York decided to put a figure on the actual number of monophagous insect species on the planet. So he took some equipment and headed for one of the richest areas of the Western Hemisphere (biologically speaking, that is): the rainforest in Panama. He selected some trees of one particular species, and placed plastic funnels underneath them. Then he enveloped each entire tree in a cloud of quick-acting, non-persistent insecticide, using a machine termed the 'bug bomb'. As the dying insects came raining down from the foliage, they were collected in the plastic funnels, after which Erwin counted and sorted his bounty.

Erwin, a beetle specialist, occupied himself with sorting out the beetles. He found 682 different herbivorous species (a staggering number in itself). He then estimated that, on average, some 20% (that is, 140 species) of his sample were associated exclusively with this type of tree. There are around 50 000 different tropical trees altogether. If each species houses a similar number of beetle species, the total number of tropical tree-feeding, monophagous beetle species must amount to somewhere around 7 million. As beetles normally account for about 40% of all insects, 17.5 million monophagous insect species in the tropics seemed a reasonable, albeit mind-boggling, estimate. Erwin's work has been both widely publicized and widely criticized, mainly because of the string of assumptions in his calculations. Nevertheless, even the lower limits of re-evaluated estimates (Kevin Gaston in 1991 arrived at 'only' 5 million species) contrast sharply with the number of different insects known today (fewer than 1 million).

Ernst Mayr is not a beetle-collector; he is an ornithologist. It is tempting to wonder if Mayr, had he been an entomologist, would ever have developed his theory of allopatric speciation, which said that new species arise when they become isolated by geographical obstacles. Would each of those tens of millions of insect species really have formed by separation due to a mountain range, a river, a glacier, or some other geographical barrier? This intuitive uncertainty certainly plays a part in the momentum that the theory of sympatric speciation (speciation in the absence of geographical isolation) has been gathering recently. The idea started very quietly, however, back in the nineteenth century, in an orchard in the Hudson River Valley.

A chronic case of Rhagoletis

In 1862, a farmer in the Hudson River Valley of New York to his dismay discovered that a new pest had arrived on his apple trees. It was a tiny maggot that burrowed in the fruits and made them rot. Not long after, a local newspaper reported on this latest horticultural nuisance, and this attracted the attention of a minister-cum-entomologist named Benjamin Walsh. Walsh had been trained in theology at the University of Cambridge, where he had been good friends with Charles Darwin. They had spent a lot of time together discussing natural history and continued to correspond after Walsh and his wife moved to the United States. There, Walsh eventually became state entomologist of Illinois and published hundreds of papers on plant-eating insects, and would probably have written many more had he not one day decided to read his mail while standing on a railway track.

In 1864, Walsh discovered that the 'apple maggot' from the Hudson River Valley was the larva of the fruit fly *Rhagoletis pomonella*. Interestingly, this insect, half the size of a housefly with attractively mottled wings and body, was known to live normally on the red fruits of the wild hawthorn tree. But because apples were not native to North America, Walsh reasoned that the fly must have recently acquired a taste for this introduced fruit. He also noted that the apple flies looked slightly different from the ones on hawthorn.

Walsh wrote to Darwin about this, and Darwin, in later printings of the *Origin*, devoted some paragraphs to it: 'In several cases', he wrote in the sixth edition, 'insects found living on different plants, have been observed by Mr. Walsh to present in their larval or mature state, or in both states, slight, though constant differences in colour, size, or in the nature of their secretions.' Walsh went even further. He excitedly speculated that when the flies first colonized apples, they diverged from their hawthorn-inhabiting ancestors and became a different species. It was one of the first theories of sympatric speciation, long before even the word 'sympatric' had been coined. But Walsh had little hard data to back up his claim.

It took exactly a century for *Rhagoletis* to surface again in the evolutionary literature. In 1963, Ernst Mayr, by then a professor at Harvard University, published his influential book *Animal*

Species and Evolution. In it, he forcefully argued that the predominant, if not the only, mode of speciation was the allopatric one. After spending a few pages on the various attempts in the late nineteenth and early twentieth centuries to prove sympatric speciation (including Walsh's), and the equally frequent rebuttals, Mayr lamented:

> One would think that it should no longer be necessary to devote much time to this topic, but past experience permits one to predict with confidence that the issue will be raised again at regular intervals. Sympatric speciation is like the Lernaean Hydra which grew two new heads whenever one of its old heads was cut off.

Little did Mayr know that, even while he was writing his book, the sympatric Hydra was raising its ugly head once again. And this time it came in the shape of his very own doctoral student, a young biologist named Guy Bush. 'I was in Mayr's class in 1961', recalls Bush, who is now at Michigan State University. 'He was working on his book and he was at that time very much opposed to the notion of sympatric speciation.' Because Bush was intrigued by the controversy, he decided to do his term paper on the subject.

While milling through the literature, Bush came across Walsh's articles on *Rhagoletis pomonella* but thought the story 'anecdotal and conjectural'. In his term paper, Bush concluded that 'there is little doubt that some form of geographical isolation is needed before reproductive isolation mechanisms can become established'. But on a cautionary note, he added: 'Conventional wisdom, however, has been proven wrong many times and it may well be that some day a good case of sympatric speciation will present itself, but I feel this is highly unlikely.' So when Bush suggested to Mayr that he do his Ph.D. thesis on *Rhagoletis*, Mayr said: 'Good idea. Let's put that case to rest once and for all.'

Not long after, Bush took his wife Dorie on a four-month field trip to gather data for his thesis. When he returned to Harvard, he took a course in sociobiology. It was given by the now-famous entomologist Edward Wilson, who covered a broad range of topics, including the evolution of sexual behaviour. At the end of Wilson's course, Bush had to write another paper and decided to do it on courtship in flies. He had been studying courtship in his *Rhagoletis* flies all summer, and in his mind still lingered the images of male flies patrolling their fruit, courting females with wing-waving displays, and their stereotyped movements when trying to mount them. 'And then the penny finally dropped. The key to the whole *Rhagoletis* story was courtship.' Bush had realized that a shift to a new host plant did not only mean a new gastronomic establishment but also a new place to procreate.

Through the stomach

So why is this important? While Bush was travelling through North America, picking flies off trees, rearing them, and watching their behaviour, on the other side of the Atlantic, University of Sussex biologist John Maynard Smith had been sitting at his desk, using pencil and paper to work out a theory of sympatric speciation. Sympatric speciation, Maynard Smith proclaimed in his 1966 article in *The American Naturalist*, would be very difficult to achieve because a preference for a new food source could never split up a population of insects as long as they kept on mating within the original population.

Suppose a gene exists in two forms, or alleles (see Chapter 1), called *a* and *b*, Maynard Smith thought. Allele *a* confers a preference for fruit A, while allele *b* makes individuals prefer fruit B. No matter how strong the preferences, if insects feeding on fruit B were equally likely to mate with insects feeding on fruit A, genes would get mixed up every generation, and there would be no progress toward the formation of new species. After all, the human population does not get split into two species because some people frequent hamburger places while others like pizza parlours.

But things might be different if sexual preferences came into play, Maynard Smith reasoned. Suppose there is a second gene, also with two alleles, say, *red* and *blue*, making the insects prefer red and blue mates, respectively. Now if one of the food-gene alleles were always to occur in the same individuals as one of the sex-gene alleles, insects preferring a particular food would always mate with other insects preferring that same food. So, for instance, if they would always bear allele *a* on the food gene and allele *blue* on the sex gene, the insects feeding on fruit A would always mate with blue mates, which also prefer fruit A. That way, there would be no danger of the food-preference genes mixing up each generation, and two separate gene pools would come into existence.

Even if such genes existed, however, the genetic jumble of recombination would be a spoilsport. During sexual reproduction, blocks of genes are exchanged, so allele *a* at the food gene and allele *blue* at the sex gene will never become permanently linked (see Chapter 1). Only in the unlikely event that the food gene and the sex gene lay right next to each other, and would always be inherited hand-in-hand, would sympatric speciation be possible, Maynard Smith argued. Either that or, of course, if the food gene and the sex gene were one and the same; that is, if preferring fruit A would also coincidentally make you prefer blue partners—a very implausible scenario, as far as Maynard Smith was concerned.

But Bush had realized that, in a way, Maynard Smith's latter scenario was precisely what was going on in *Rhagoletis*. Because of the flies' habit of mating and laying eggs *on* their preferred fruit, mating was always with one particular type of mates: not blue flies, but ones that shared the same food preference. So, under these conditions, perhaps sympatric speciation was possible after all. 'I've spent the rest of my life trying to prove it', Bush proclaims proudly.

Even though Bush had cracked the problem, it was still going to be hard to convince his PhD committee. And being at Harvard and in Mayr's vicinity didn't really help, Bush says: 'I was caught, as they say, between a rock and a hard place. Most

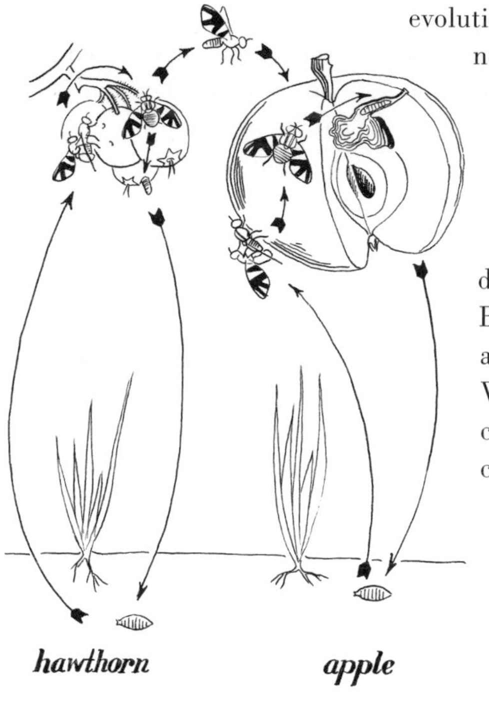

After the hawthorn fly, *Rhagoletis*, moved to apples, genes (including those for apple adaptation) were not taken back to the hawthorn population, because those flies mate only on the host plant itself.

evolutionarily inclined cognoscenti . . . viewed sympatric speciation as little better than an heretical, unsupported fantasy.' As the date of his thesis defence drew nearer, Bush grew ever more apprehensive. Whenever he got the chance, Mayr made it clear that he viewed all aficionados of sympatric speciation as crackpots. And what Bush was saying in his thesis was exactly what Mayr had hoped he would disprove: that the apple race of *Rhagoletis pomonella* (which by the 1960s had spread over most of the northern United States and was posing a serious agricultural threat) had recently shifted from its native hawthorn and was on the way to becoming a separate species.

One evening in January 1964 Bush was putting the finishing touches to his controversial thesis. Mayr would be leaving on his vacation in a few days, and he had set Bush a deadline for delivering his draft that very evening. But it was already late. As Bush approached Mayr's home he noticed that the lights were off. 'He had told me that if his porch light was out it would be too late. On the outside chance that he had forgotten to turn the light on,

I rang the doorbell anyway.' Mayr informed him that it was indeed too late and that Bush better find someone to replace him on the Ph.D. committee. 'I left with mixed feelings', says Bush, who does not think Mayr was deliberately avoiding him. 'Inwardly I breathed a sigh of relief while at the same moment regretting he would not be on my committee or read my thesis.'

Bush found a replacement and graduated without any problems. He then moved to Melbourne, Australia, to further his studies, and stayed there for two years. In 1966, he returned to Harvard to prepare his thesis for publication and—that harrowing initiation of every scientist—to present his results at a scientific meeting. After Bush delivered his talk, Mayr's buddy Theodosius Dobzhansky is said to have stood up and commented: 'That's very interesting, but I don't believe it. Sympatric speciation is like the measles. Everybody gets it, but they all get over it.' But Bush didn't get over it. In fact, he seems to have been the source of a complete epidemic. But, first, let us have a look at Bush's reasons for going against orthodoxy.

The miracle in the apple tree

As Walsh had noticed a century earlier, the apple–hawthorn situation simply didn't look like it had come about by geographical isolation. The two types of tree grow shoulder to shoulder throughout the Hudson Valley. And given farmers' paranoia about insect infestation, it is unlikely that the apple fly had been quietly evolving in some 'allopatric' apple orchard somewhere else. If that had been the case, the unfortunate apple farmer would immediately have notified Walsh or one of his colleagues. So sympatry, yes. But speciation? Perhaps no. After all, what evidence was there that apple and hawthorn flies were really different? For all Bush knew, hawthorn flies had simply broadened their taste to include an introduced fruit. What he had to do was show that the two were genetically different.

Those were the days before protein electrophoresis (not to mention DNA fingerprinting and the like), so Bush had to resort to examining the insects' shape—their morphology. Eventually, he zoomed in on a particular difference in the fly's ovipositor. This sting-like organ, which the female uses to inject her eggs under the peel of the fruit, was exactly 1.2 millimetres long in all apple flies. But in the hawthorn flies, the excited Bush found it varied in length from a puny 0.9 to a whopping 1.5 millimetres.

Now if the two host-races were interbreeding continuously, Bush reasoned, one would expect all flies to have the same variability. The fact that this was not the case he took as evidence that the two races were somehow genetically different. Possibly, the apple fly had arisen from only a small number of closely related ancestors, which all had had similar-sized ovipositors. And that would tally with the observation that it had first been recorded in a small area near New York and had since spread eastwards, across the Great Lakes into Massachusetts and Connecticut, southwards to North Carolina, and northwards into Canada.

In 1975, Bush wrote up a more detailed scenario of what he thought went on in sympatric speciation by host-shift. This is what he envisaged. Imagine a fly specialized for eating hawthorn. One fine day, a few mutant flies are born with an innate preference for something more juicy, something less like those boring small hawthorn berries. Something more like those big red things hanging from that tree yonder. Responding to the call of nature, the mutant flies move to the apple tree, breed there, and produce more flies with their own taste preferences. Because the entire life cycle—mating, the laying of eggs, and the development of the larvae—took place on one of the two host fruits, genetic separation would come naturally. That way, genes would circulate only among the flies on a particular food plant.

But Bush's speciation scenario was not completely watertight. The two host-races might be isolated to a certain degree, but nothing would prevent the occasional trespasser from the hawthorn clan mixing with the apple flies, and vice versa. A more definitive separation would be necessary to complete speciation. The key was timing, as Bush's colleagues Catherine and Maurice Tauber from Cornell University in New York noticed. If the host plants differed in the time of year that their fruits ripened, this would further reduce the chances of the two host-races meeting.

So all these factors—the appearance of mutants, the link to the host plant, and a difference in fruiting time between the two host plants—would have to be in place for speciation to occur. At first glance, the *Rhagoletis pomonella* host-races seemed to fit the bill exactly. A *Rhagoletis* male would set up camp on a fruit, and mate with females that arrived there. The females would then lay their eggs on the same fruit, and the larvae would develop and hatch in it. On top of that, Bush had found that there was a lag between the emergence times of the apple race and the hawthorn race. Most apple flies emerged around 25 July whereas the hawthorn flies came out three weeks later. Normally, therefore, most apple flies would never encounter a hawthorn fly. And then, of course, there was the difference in ovipositor length, which seemed to indicate that the races were already genetically different.

By the late 1970s, then, Bush seemed to have a good case for saying that the apple race of *Rhagoletis pomonella* was well under way to becoming a separate species, even though it had never been geographically isolated from its hawthorn-inhabiting ancestors. Bush's work began to be cited in textbooks as a compelling example of sympatric speciation. But not everybody was convinced that Bush's case was established beyond doubt. Douglas Futuyma, an evolutionary biologist at the State University of New York at Stony Brook, recalls: 'Guy Bush at the time had a lot of intuition, but very little evidence.'

In a 1980 article in the journal *Systematic Zoology*, Futuyma and his student Gregory Mayer showed that the *Rhagoletis* story was based on—to say the least—shaky evidence. Most essential data appeared to be missing. For example, it was not known whether the choice that females made for the fruit to lay their eggs on was genetically based. This was crucial because, for evolution to work, inheritance is necessary. If the offspring of apple flies returned to hawthorn to lay their eggs, and vice versa, speciation could never happen. Experiments to investigate this would have been easy, Futuyma and Mayer argued. In fact, they claimed, ecological work done in the late 1970s seemed to show that flies hatched from apples would lay eggs on the 'wrong' fruit as easily as on their host plant.

The following year, Futuyma's colleague John Jaenike independently wrote a similar critique for *The American Naturalist*. Were the two host-races really different?, he wondered. Bush had tried to show this with his ovipositor data, but Jaenike pointed out that the sizes of the flies on apple and hawthorn were also different, which could simply be the result of better or worse food quality, or climate. So if the ovipositor differences were just a reflection of differences in size, they could not be used as evidence for a genetic gap between the two races. On top of that, Bush had always stressed the different time of the year that the flies of apple and hawthorn emerged, which was adapted to the time of fruiting of the hosts. But Jaenike exposed the circularity in this reasoning. Because the host-races were identified only by the host they lived on, said Jaenike, and the fruiting times of the hosts differed, there was no proof of anything in Bush's observations: there were just flies emerging through the entire summer, with early-emerging flies finding apples to lay their eggs on, and late-emerging flies having to make do with hawthorn.

It seemed a shattering critique, and to many biologists the footing of the *Rhagoletis* study was firm no longer; it had crumbled. But Bush was undaunted. 'At first it made me mad', he admitted in 1998. 'But just when Futuyma and Mayer's paper came out, I was moving to Michigan State University, where I had accepted a professor's position. There, I had access to large laboratory facilities, including a molecular biology lab.'

Counter-offensive

So rather than starting a bitter trench war, or admitting defeat, Bush realized that the burden of proof was on him. He joined forces with entomologists from other universities, and hired a bunch of Ph.D. students. One by one, they started tackling the major questions that remained.

David Courtney Smith, from the University of Utah, had a better look at the times that the flies took to emerge from their fruits. He collected infested fruits and brought them to the laboratory for the flies to emerge. He then allowed the apple flies to mate among themselves, and the hawthorn flies to do the same. The eggs these flies laid were then reared under identical temperatures and on an artificial diet. Nevertheless, the apple flies of the next generation still appeared some 18 days earlier than did the hawthorn flies. So in contrast to what John Jaenike had suspected, the difference in emergence time really was genetic. This implied that the two host-races had fine-tuned their development to the fruits' seasonal rhythm. Jeffrey Feder, one of the grad students in Bush's molecular biology lab, studied the genes of apple versus hawthorn flies, and found that the two host-races carry substantial differences at three of their chromosomes. Virtually the same result was obtained by Bruce McPheron and Stewart Berlocher, from the University of Illinois. So the ovipositor had been only the tip of the iceberg: the two host-races did differ genetically after all.

The three papers by Courtney Smith, Feder, and McPheron/Berlocher appeared as an impressive triple bill in the 3 November 1988 edition of *Nature*. A commentary in the same issue suggested that 'perhaps the patron saint of evolution is, after all, sympatric'. There were more supportive results to come out of the labs of Bush and co-workers. The University of Massachusetts entomologist Ron Prokopy countered Futuyma's objections by discovering that the offspring of an apple fly are much more likely to lay their eggs on apples than hawthorn flies.

Apple-reared males show something similar: they tend to land on apples more frequently than do their hawthorn-reared relatives.

But in 1994, Feder (by that time an assistant professor at the University of Notre Dame) showed that in spite of the genetic differences between the two races, they continue to mate and exchange genes. 'After the fruit have fallen, the flies pupate in the soil, so we put nets over the ground below both types of tree to catch all the flies the next summer', he said in an interview.

> We then tagged some 10 000 of them with coloured liquid paper marks and released them. After a few days we went along all the trees in our field to check how many apple flies had landed on hawthorn fruit and vice versa. Overall, we found that some 6% of all flies had made the wrong choice.

This 6% level of 'fruit infidelity' was higher than expected. Feder continued:

> With such a level of genetic exchange between the apple and the hawthorn population, the differences that we see should have long disappeared. The two races should fuse. Apparently, some sort of natural selection is keeping the two races apart.

In other words, apple flies that try their luck at hawthorn apparently do not fare as well as born and bred hawthorn flies, and vice versa. This phenomenon is known as a fitness trade-off.

But people had already looked for such fitness trade-offs to no avail. In 1988, Prokopy had found that it was *not* true that apple flies survive best on apple and hawthorn flies survive best on hawthorn. If anything, hawthorn is always a better host, regardless of the type of flies. So where does the selection take place? 'It now seems that it is not just the quality of the fruit itself that matters, but the fruiting phenology', says Feder. Cryptic words indeed. More simply put: apples fall from the tree some two to three weeks earlier than do hawthorn fruits. This means that apple maggots that crawl out of the fruit and burrow in the ground experience a bit more of late summer warmth than hawthorn maggots normally do. According to Feder, this is fatal in hawthorn maggots because it upsets their development. 'Normally, they go into diapause, a lengthy resting stage before they pupate and emerge the next summer. But when they get these high temperatures, they often skip their diapause and go straight to adult development, which means they emerge just before winter and die.'

So Feder suspected that apple maggots have to go into deeper diapauses to circumvent this problem. He and his team tested this idea by doing an elegant experiment: they put hawthorn maggots developing on hawthorn fruit in different incubators in the lab. One incubator was set to mirror the temperatures normally experienced by the maggots while the other was set to have a longer period of warm weather before the onset of winter: it artificially mimicked the temperatures experienced by apple maggots. The results, published in 1997 in the *Proceedings of the National Academy of Sciences USA*, were astounding. Hawthorn flies from the incubator simulating the apple maggot temperature regime actually became genetically more like apple flies. What had happened? Apparently, whatever genes make an apple fly are already present in the hawthorn ones. In Feder's experiment, only the hawthorn flies with a genetic make-up that was suitable for surviving the pre-winter warm spell had made it. That genetic make-up is exactly the one that characterizes the apple race. It was a rare insight in the subtle workings of sympatric speciation.

Joining up the dots—once again

So, rather than 'putting the case to rest once and for all', as Mayr had intended, the project Guy Bush started almost 40 years ago has now grown to an impressive mountain of evidence in favour of sympatric speciation. 'I'm delighted by what they've found', says Futuyma generously. 'Bush and Feder and all the others have really made a good case. I'd say the burden of proof is now on the sceptics.'

Not that many disbelievers remain. Even Ernst Mayr seems to be convinced of the validity of the *Rhagoletis* studies. 'It would seem that Guy Bush was indeed right', Mayr admits. But he still adds cautiously: 'This is somewhat . . . what shall I say . . . hypothetical, but that is what it is beginning to look like.' Most other biologists have changed their minds too, as developed during a symposium held in Bush's honour in Asilomar, California, in 1996, intriguingly entitled 'Endless Forms'. Most delegates were persuaded then that sympatric speciation can and does happen.

But major questions remain. To begin with, in spite of the rapid evolution that appears to have taken place in *Rhagoletis* in little more than a century, the two host-races are still not completely separated. Would or could the two host-races of *Rhagoletis pomonella* eventually acquire full species status? And, if so, could *Rhagoletis* still be a freak of nature, or is similar speciation rampant in insects and other animals? Does it explain the bewildering diversity of tropical herbivorous insects?

As to the first question, the experts are divided. 'I think speciation in *Rhagoletis pomonella* is still proceeding', says Bush. Feder is more careful: 'It could very well be a stable situation, which does not proceed any further. The fly is in a sort of speciation limbo: it has gone part of the way, but the two hosts are not different enough.' The problem is that apple and hawthorn fruiting times are separated by less than three weeks, which means that the two races will always overlap somewhat. If the shift had been to a host that fruits, say, six weeks earlier than hawthorn, separation could have been complete from the start.

That is why Feder, Berlocher, and their co-workers have now turned their attention towards other species of *Rhagoletis*. Many species closely related to *pomonella* live on very different host plants. *Rhagoletis mendax*, for example, lives exclusively on blueberries. And yet, *pomonella* and *mendax* can be crossed in the laboratory almost as easily as if they belonged to the same species. What keeps them separate species in nature seems to be solely the adaptation to different host fruits. So in any case, the things that are responsible for the separation of full species in *Rhagoletis* are the same as those that keep host-races apart. Stewart Berlocher has recently discovered a couple of new flies on dogwood and sparkleberry that seem to come somewhere between the host-races of *pomonella* and the full species of *mendax*. Both are more genetically different from *pomonella* than the apple and hawthorn races are from each other.

'I like to compare it with Darwin's theory of coral-reef formation', says Feder, referring to Darwin's 1842 book *The Structure and Distribution of Coral Reefs*. In this book, Darwin deduced that coral reefs, contrary to the prevailing opinion at the time, are not sitting on top of extinct volcanoes but form by growing upwards while the sea bottom sinks. Darwin reached this conclusion by comparing coral reefs at various stages in their growth. 'With *Rhagoletis* we're doing essentially the same', muses Feder. 'We compare situations where two interbreeding host-races exist, like in *R. pomonella*, with those where separate species have formed, and then try to connect the dots.' Ironically, Ernst Mayr used the same approach to find support for allopatric speciation (see Chapter 2).

Even if the dots can be joined up, however, it still doesn't mean that an apple–hawthorn-like situation will necessarily move to a *pomonella–mendax* situation. You are still stuck with that 6% gene flow and it is hard to see how and why that should be reduced to zero gene flow. 'I keep on wondering, gee! How do you go through the transitions?', Feder admits. But he has one possible solution: stepping stones. 'Possibly,' says Feder, 'yet another host-race can spring from the apple race. For example, at the beginning of the season, before apples have fruited, some adventurous flies might venture into a tree that fruits even earlier than apple. That way, you could have host-race formation all over again, this time on a fruit that is temporally completely isolated from the original hawthorn.' And Bush has, in fact, observed just such a situation in a cherry-infesting form of *pomonella* that he stumbled on in Wisconsin.

Of course, this is no more than a seasonal variant of the ring species that we've seen in Chapter 2. Even though the hawthorn and the new host-race are isolated, they can still exchange genes via the intermediate apple maggot. Feder concedes that many intermediate situations can be envisaged. But that is just how nature appears to work. Species and their origins are not always clear-cut. 'The very fact that it's sloppy is what makes it all so interesting', he says.

CHAPTER SEVEN

A Freak Show?

Apple Maggots are not Alone

The story of the apple maggot fly has reached stardom in scientific circles as the first good example of speciation taking place without any geographical barriers. Even Ernst Mayr and other opponents now have to admit that what happened in that Hudson Valley orchard 200 years ago bears all the hallmarks of sympatric speciation.

But the *Rhagoletis* story may be a bit exceptional. The fly just happened to be in the right place at the right time. After all, it is not every day that a fruit-eating insect is treated to a new, large and juicy agricultural product; especially a fruit that can be enjoyed alone—because no other fruit-eating insect has discovered it. A fruit also, where parasites and predators may be absent, because they, too, have yet to learn to patrol the new, unfamiliar tree for possible prey. In other words, isn't what happened in *Rhagoletis* just a freak, a one-off? An interesting, but unique exception to the rule of allopatric speciation, possibly spawned by the human habit of moving food plants around the world, something that otherwise would not have happened on such a scale?

Ernst Mayr certainly would like it that way. Over the phone he says of Guy Bush, his former pupil, 'Now *he* went much too far. He thought that all speciation in insects was sympatric speciation. But there is no evidence for that.' Granted, perhaps not all insect speciation. But there are many more situations that suspiciously resemble the *Rhagoletis pomonella* affair, both in artificial and in natural environments. None has such good credentials as *Rhagoletis*, but many have been studied well enough to make them prime candidates.

First of all there are agricultural pests that, like *Rhagoletis*, seem to have gone into an initial stage of speciation after a new agricultural crop was introduced in a region. One example involves the codling moth *Cydia pomonella*. This small, grey-and-brown insect has always been infamous in Europe for causing 'wormy' apples and pears. The caterpillars burrow in them, surprising the unwary apple-eater with holes, frass, and bisected larval bodies. In the middle of the eighteenth century, codling moths were accidentally introduced into North America, where they spread through apple and pear orchards across the continent, reaching the west coast in 1873.

Until the beginning of the twentieth century, the codling moth stuck to its natural hosts: apples and pears. But, in 1912, it was discovered attacking walnut trees that grew around pear-packing sheds in California. From there, the walnut infestation spread through most of the state. In the 1930s, starting from a walnut grove near Los Angeles, yet another host-race appeared, this time on plums. The plum race subsequently spread all over the state and became a nasty pest in its own right. Entomologists who studied the codling moths later discovered that the ones from walnut preferred walnut over apple to lay their eggs on, even if they had been reared on apples. So, paradoxically, did the moths from plum, which confirmed the suspicion that the plum race had recently sprung from a walnut-infesting stock. The plum race developed faster than either the walnut or the apple race, which matched the fact that plums mature earlier than do the other two fruits. There was also yet another adaptation of the different host-races to the seasonality of their respective fruits: caterpillars of each host-race went into diapause at different times of the year. The plum race began in June, the walnut race in mid-July, and the apple race, finally, in late July.

So, although the information on codling moths is more sketchy than that on the apple maggot fly, it, too, seems to have evolved into new host-races over a short timespan, apparently taking the first strides toward speciation, and possibly following the stepping-stone scenario envisaged by Jeffrey Feder for *Rhagoletis*. Like the apple maggot fly, the codling moths have sequentially acquired egg-laying preferences and life histories that suit their new hosts, and at least some of those characteristics are genetic.

We could go on. Agricultural entomologists are always on the lookout for new and possibly harmful pests. Their freak show now features several dozen recently evolved host-races of native insects on introduced plants—and the reverse, introduced insects on native plants. Step right up! The programme includes a moth miraculously splitting into five host-races on as many introduced banana varieties on Hawaii; the amazing native moths on the Pacific island of Rapa, which only live on an introduced weed; and the incredible spider mites (this time not insects but relatives of ticks and spiders) splitting into cucumber- and tomato-infesting forms in greenhouses.

At least one thing stands out from these agricultural examples: host-races tend to evolve very quickly when a new host becomes available. The changes in the apple maggot fly took place in some 300 years and just as many fly generations. The codling moth evolved into the walnut race between 1912 and the mid-1970s, which probably amounted to some 200 moth generations. Apparently such short times may be enough to adapt physiologically to the new host and get the initial separation under way. If this is the case, then it is not surprising that the exciting intermediate stages are normally encountered only in agriculture, where new hosts or new pests are introduced and new host-races, if they appear, are noticed and dealt with quickly because of the economic importance. In nature, host shifts will largely go unnoticed and intermediate stages will be rare if speciation is swift.

Caught in the act

But sometimes luck lends a hand. For twenty years now, Steph Menken of the University of Amsterdam has been studying moths called *Yponomeuta*. These tiny moths have snow-white wings, speckled with black dots, as if they were wearing miniature ermine capes. Indeed, they are commonly known as 'small ermine moths'. But it is not their regal appearance that attracted Menken. 'They are good candidates for sympatric speciation', he claims. 'Like the apple maggot fly, they mate on the host, and they are very fussy about their food plant.' Most of the 70 or so small ermine moths specialize on species from the spindle-tree family. But many of the European *Yponomeuta* have moved away from the traditional hosts and started feeding on willow, poplar, members of the rose family, and other plants.

One of these unorthodox Europeans is *Yponomeuta padellus*, a species that has an unusually broad range of food plants. In the Netherlands, it is found on hawthorn, blackthorn, wild plum, cherry plum, and juneberry. In Scandinavia, it also occurs on rowan trees. Menken's colleague Léon Raijmann says: 'It was certain that rowan trees, which are very common in the Netherlands, had always been free of these ermine moths'. That is, until 1985, when a small population was discovered in the province of Gelderland. Then, seven years later, a colony on rowan suddenly appeared along the west coast of the country, some 200 kilometres from Gelderland. From this new locality, the rowan-feeding habit spread massively and within a few years *padellus* all over the Netherlands had broadened their taste to include rowan trees. 'At first we assumed that the Dutch populations on rowan were invaders from the Scandinavian populations', says Raijmann. But when the researchers looked at genetic markers, it immediately became obvious that they were not. The Dutch rowan-feeding moths were direct descendants of the populations feeding on other food plants nearby.

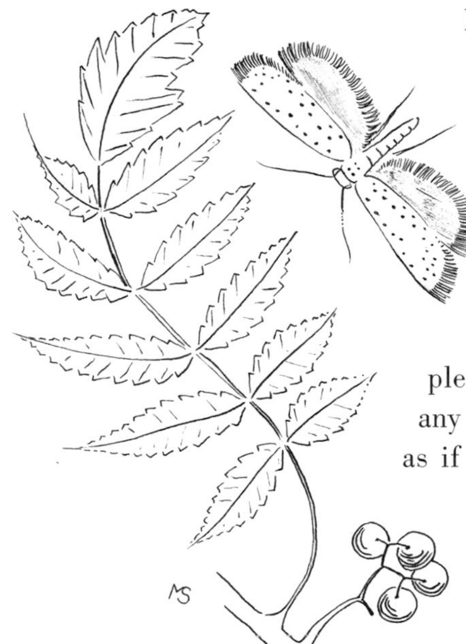

A small ermine moth, *Yponomeuta padellus*, on rowan.

Rinny Kooi of Leiden University, meanwhile, found out that the rowan-feeding moths would also lay their eggs on other food plants, but that the reverse was not true: moths from, for example, hawthorn refused to lay any eggs on rowan. So it looks as if the Dutch entomologists had been witnesses to a rarely observed event: the evolution of a new host-race under natural circumstances. Raijmann has some notion of why this shift could have taken place so quickly and so massively. 'Normally,' he says, 'small ermine moths suffer quite heavily from attack by parasitic wasps, which lay their eggs in the caterpillars. But the 1985 population on rowan in Gelderland was virtually free from parasites, whereas neighbouring populations on other host plants were under heavy attack. In fact, the populations on rowan have remained free of parasitic wasps ever since.'

So *Rhagoletis* is not a freak of nature, it seems. It enjoys the good company of codling moths, spider mites, small ermine moths, and probably also treehoppers, goldenrod flies, leaf beetles, and lacewings. Insects everywhere, harmless meadow-dwellers as well as noxious agricultural pests, regularly make the first strides towards speciation. Or at least that is what many entomologists are beginning to believe. What's more, speciation due to host-shift can apparently be a rapid process. Exactly what steps the insects take, however, and in which sequence, remains mysterious.

A matter of taste

The first mysterious detail lies right at the start of the host-shift. When Guy Bush, the chief proponent of sympatric speciation in the apple maggot fly (see Chapter 6), presented a detailed version of his speciation scenario, he began by suggesting that first a new mutant appears that can feed on another fruit. This is obviously a key feature, but exactly how should we envisage it? What sorts of genes are responsible for food preference? Could it be a single gene or does it require many genetic changes? Or is the necessary genetic make-up already present in the ancestral population, only waiting to be released by the appearance of a new fruit?

The last possibility—that the genetic make-up is already there—may sometimes be the case, as it seemed to be in the hawthorn *Rhagoletis*, which can be turned into something resembling an apple fly simply by forcing it to develop under apple circumstances. But at least in the situation of the rowan-feeding small ermine moths above, it probably wasn't. After all, rowan had remained untouched for decades, until suddenly a population of rowan-feeding moths appeared. To answer these questions, therefore, it will be necessary to find the genes that code for insects' gastronomic predispositions. Bush says: 'My main ambition at this moment is to pull out the genes that are responsible for host selection.' Steph Menken has similar plans with his *Yponomeuta* moths. He is dying to find out what factors of the host plants are picked up by which sense organs—and from that he hopes to find the underlying genes responsible. 'We are now working from the outside to the inside', Menken explains.

A moth has smell and taste receptors in so-called 'sensilla' on its antennae, legs, and ovipositor. When these are stimulated by odour or taste molecules, they send out weak electrical currents to the insect's central nervous system. Menken's colleague Peter Roessingh is using microscopic electrodes to tap the electrical currents from a single receptor to see which smells trigger a response, and which do not. Finding out how receptors in invertebrates work is still in its infancy. But the genetic basis for change can be very simple. For example, a change in a single letter in the DNA code of the roundworm *Caenorhabditis elegans* can put the animal off a substance called diacetyl, which they normally are attracted to, says Menken. And a similar mutation causes a strain of *Drosophila* to go for salty rather than sugary substances. It is likely that similar mutations cause plant-feeding insects to acquire a taste for a different food plant. Menken summarized the problems:

> We do not know how often such mutations occur in monophagous insects and what effects they have. Recording electrical pulses from insect sensilla has been going on for some time. But it is laborious and painstaking work and usually researchers are satisfied with studying only a few, often related individuals. Worse, they tend to discard test animals that 'do not respond typically', while these may in fact be the very mutants that we're interested in.

Let us assume that the required genetic mutations pop up regularly in a population. This does not mean that each mutant will find a new host or that it will successfully colonize it. Even if a food plant to their taste is available, it may already be occupied by some other insect. Or there may be too many predators or parasites on it. The absence of competition and enemies will help make the colonization a success, just as it has in the rowan-feeding *Yponomeuta*. Colonization is one thing, sticking it out is another. Even if the founders carried some alleles that made them prefer the new host, it is unlikely that all their offspring carry these alleles. The genetic jumble that takes place during reproduction will leave some offspring with only ancestral alleles. So what would prevent these from mingling with their cousins on the good old host?

There are two things that would. One is nature, the other is nurture, and both are probably important in sympatric speciation. The 'nature' factor is the habit of many parasitic insects to have sex on the host, rather than to disperse, mate, and then find a host. We've seen this in the apple maggot fly, and it happens in small ermine moths and in lots of other insects as well. In fact, it is probably one of the key features of sympatric speciation because it means that the genes that are needed for specialization on the new host will remain there and not be lost by mixing with other populations.

The so-called Hopkins effect is the 'nurture' thing that may aid in the colonization of the new host. Named after the entomologist who first described it in 1917, the Hopkins effect is the phenomenon that insects that have (as larvae) fed on a certain type of food will (as adults) prefer that food to lay their eggs on. So this is a learning thing, which may nevertheless consolidate the link between the new host and its mutant occupants. The Hopkins effect is not rare: it has been found in many different kinds of insects, including small ermine moths and the apple maggot fly.

A food-preference mutation and sex on the host are probably the two most important factors needed to get the initial stages under way. This opens the road to increased specialization. The new host will probably not be identical to the original one in many respects: it will have different toxins to which the insects need to evolve tolerance. It will have a different rate of ripening and rotting, which will require shifted life cycles in the insects. The surface of the leaves or bark may have a different colour and texture, and this may mean the insects will evolve different camouflage patterns, different ovipositors, different jaws, and so on.

All these bits of microevolution can occur slowly, surely, and simultaneously, once the two populations have been separated by the forces of food preference and sex-on-the-host. As the two populations go their own separate ways they will diverge further and further and sooner or later qualify as different species. But hang on! This may be so, but we have seen in Chapter 1 that an important (though perhaps not decisive) feature of species is that they are reproductively isolated: they usually do not cross-breed. Now, even if the new host-races mate on the host, accidents are bound to happen, especially when the new host grows close to the old one. In the apple maggot fly, we have seen that some 6% of the fruit flies end up settling in the wrong place. And their offspring are no less viable or fertile than if they had mated within their own host-race. So how do all those other changes take place?

The fact is, along with the natural selection acting to adapt to the new host, there is an undercurrent of other genetic changes, happening purely accidentally, which will over time increase the reproductive isolation of the two host-races. One way this can happen is by the curious phenomenon of pleiotropy.

Synchronized sex

Pleiotropy (from the Greek *pleios*, 'more', and *trope*, 'change') is a term used to indicate that genes may have more than a single function. Hence, if they change, more than one characteristic may be altered. A Japanese entomologist named Takahisa Miyatake stumbled across such a pleiotropic effect that may be important in speciation. Miyatake, who works at the Okinawa Prefectural Agricultural Experimental Station, made his discovery while studying the melon fly *Bactrocera cucurbitae*, a notorious pest of melons, cucumber, and other members of the gourd family.

By breeding only from flies that emerged from their pupae very early or very late Miyatake produced after about 45 generations a 'fast' and a 'slow' group; the former took only 6 days for its larval development, the latter more than 12 days. To his surprise, Miyatake noticed that along with the change in developmental time, the flies had also experienced a jolt to another biological clock. In nature, mating takes place around dusk. But in the fast developers, sexual activity peaked in late afternoon, one hour before dusk, whereas the slow developers started

Pleiotropy: when the life cycle is shortened, the timing of mating also changes (at least in melon flies).

getting interested in sex only some three-and-a-half hours after nightfall. According to Miyatake, both rhythms seem to be regulated by the same clock-gene.

To see how this would influence the ability of the slow and the fast lines to cross-breed, Miyatake and his student Toru Shimizu marked flies from both lines, put them together in cages, and watched to see which flies mated with which. As expected, most matings took place between males and females from the same line. Animals from the fast lines would hardly ever mate with those from the slow line, and vice versa. Apparently, melon flies are like humans: if your partner has a different biological clock your sex life suffers.

Pleiotropic effects of this sort, where development and the timing of sexual activity are genetically linked, could be important in sympatric speciation. For example, when an adventurous population of flies starts feeding on a new fruit that rots quickly, this would be to the advantage of fast-developing larvae. So natural selection will cause the flies feeding on this new fruit to condense their development into a shorter time period. If mating time then also changes, this could help prevent the adventurous flies from cross-breeding with the ancestral stock. In fact, some flies related to the melon fly differ from each other in mating time. An important difference between *Bactrocera tryoni* and *B. neohumeralis*, for example, is the fact that the former mates at dusk, the latter in the daytime. Other groups of insects also have agendas that do not overlap. Each of the 69 North American giant silkmoth species, for instance, has its own time-window for mating: 4–6 p.m.: *Callosamia promethea*; 3–4 a.m.: *Hyalophora cecropia*; and so on.

We now have brought together all the pieces that make up sympatric speciation: a vacant ecological niche; a genetic predisposition that makes some individuals seek it out; and then rapid adaptation to the new niche. Given that the new host-race will have to juggle its genes to adapt to many different features of the new environment at the same time, pleiotropy will cause a whole suit of genes to change as well, including many involved in reproduction. As Bill Rice (long-time bottleneck critic from Chapter 3 and the breeder of the female-unfriendly banana flies from Chapter 4), has remarked, sympatric speciation is 'simple and it is obvious'.

So far, we have implicitly equated sympatric speciation with host-shift speciation in insects. The best examples are from this species-rich group—an estimated 17.5 million species worldwide. It is perhaps logical to focus on plant-eating insects first, as we have here, but there is no reason why sympatric speciation could not occur under other circumstances, and among other groups. Replace 'host' by 'habitat', submerge it all in fresh water, and the stage is set for a group of organisms that are the next big thing in sympatric speciation research: freshwater fish.

Fishy business

The Zoology Institute of the University of Munich is housed in a grey building along the tree-lined Luisenstrasse. On the second floor, in a laboratory stuffed to the brim with equipment, computers, and jars with fish in formalin, one finds graduate student Uli Schliewen. 'Since I was a schoolboy, I have been fascinated by mouthbrooding cichlids', Schliewen muses, pointing at the perch-like fish in spirit. Looking at the goldfish-sized animals hanging grey and stiff in their jars, it is hard to imagine them busy with their intricate and bizarre courtship. Directly after the female releases her eggs in the water and the male envelopes them in a cloud of sperm, the female turns around and scoops up all the fertilized eggs in her mouth. She then fasts for weeks, while her babies grow in her mouth. In some species, the male even sacrifices a bit of his own mouth-space to help out.

While still in secondary school, Schliewen heard that a professor Hans Peters at the University of Tübingen had been studying mouthbrooding cichlids from crater lakes in the West African state of Cameroon. He asked Peters if he could come over to Tübingen to see them.

> He invited me for a whole week, and after that visit, I was hooked. The image of clear crater lakes surrounded by rainforests somehow nourished my tendency for escapism. And I realized that the fish were special, because they had such different breeding habits: the fish in one lake were mouthbrooders, the ones in another were not. I was not really thinking of speciation yet at that time.

In 1989, when Schliewen was a biology student at the University of Munich, he was determined to study the breeding behaviour of the Cameroon cichlids. So he arranged to do his master's thesis on the subject at the nearby Max-Planck Institute for Behavioral Physiology in Seewiesen. The next hurdle was getting the finance sorted out to pay for his African expedition. Help came from a cornflour-producing company, which gave him a modest travel grant. Schliewen drags a topographic map from a shelf and puts his finger on a tiny pale-blue spot, one of the 40 waterlogged relicts of ancient volcanoes in the African state. 'This is Bermin,' he says, 'a minute lake half a square kilometre in size.' Getting there proved harder than Schliewen had expected. When he reached the area, the local Bakossi people would not allow him to see it, because Lake Bermin is a holy place to them and is said to house the Mamiwata, a waterborn spirit of ambiguous nature. 'I first had to explain to them why I wanted to go diving there, which took an entire day and involved a sacrificial ceremony', Schliewen says. The villagers finally yielded, but warned him that he would never be able to reach the bottom, because it was a lake with 'no end'. Lucky for Schliewen, whose cichlid fish are bottom-breeders, 'no end' meant 16 metres. The lake was so remote that he could not have carried any heavy scuba equipment there, which meant all exploration had to be done by snorkelling (quite a spartan undertaking in the deep water).

During the several weeks that Schliewen spent exploring every nook and cranny of the lake, he came across a flock consisting of nine different species of cichlids. All were tilapias, which are well known among aquarists and gastronomists alike, but some had characteristics that are not so typical for this sort of fish. For example, one of the Bermin species specializes on eating freshwater sponges, another lives on plankton, while the remaining seven feed on vegetable debris, or detritus, on the lake bottom. Among the detritus feeders there were other differences. One tiny species of less than 4 centimetres long, named *Tilapia snyderae* by Schliewen, frequented shallow water along the shore, while the large *T. bythobates* (three times as long and some 20 times as heavy) lurked in the deep, where it had apparently found a way to cope with the oxygen-poor environment. Schliewen also noticed a variety in body shapes and strikingly different breeding behaviours. The yellow and elegant *T. flava* formed breeding colonies with females tending each other's nests, while pairs of the robust, thick-lipped *T. bakossiorum* made their nests far away from any neighbours.

With so many related fish species mingling in such a small lake, mixed couples would be expected. And yet, although he looked hard, Schliewen never saw a mating pair of two different species, nor did he ever manage to find a single intermediate fish. So cross-breeding was apparently very rare. The strange thing is that none of these cichlids occurs anywhere else: they are endemic to lake Bermin. Moreover, the little lake is practically isolated from the outside world. Water trickles away from it only via a small waterfall that cichlids cannot possibly ascend. So where did all those fish species come from? To answer these questions, Schliewen realized that he needed to delve deeper into the evolutionary history of his favourite fish. By now a graduate student, and still funded by the Max Planck Society, he sought cooperation with Diethard Tautz, a molecular biologist turned evolutionary. Bringing together Tautz's expertise with his own love for fish proved a fruitful marriage. The two spent a few years pulling DNA sequences out of the African fish and Schliewen can now say with confidence:

They evolved there on the spot from a single ancestor which probably got there via a creek; we've looked at the DNA of the fish and it turns out that they are all extremely closely related. Yet they are all different species. Not only do they look different and they occupy different niches, they also avoid cross-breeding with each other. Sympatric speciation. *Zweifellos*!

So how did it happen? Schliewen thinks that the empty lake offered a range of ecological 'niches' to the fish, which allowed them to diversify to exploit these niches more efficiently. For example, Schliewen says, if there were large and small food items, specialists would suffer the least from competition. This would mean that exceptionally large and exceptionally small fishes would be at an advantage, because they would not compete so severely. Here pleiotropy comes into play again, because small fish tend to mate with other small fish, and likewise for large fish. This produces reproductive isolation, and the whole thing could spiral out of control to produce two separate species, one large one specialized for eating large chunks of debris, the other smaller, adapted to small bits.

The presence of a cichlid species flock turned out not to be restricted to Lake Bermin. Another tiny crater lake in Cameroon, called Barombi-Mbo, has a similarly unique fish fauna: here, too, a bunch of different cichlids swim around that must all have arisen from a single stock recently, as they have very similar DNA. Schliewen and Tautz published their results in 1994 in the journal *Nature* as new evidence for sympatric speciation. Tautz comments: 'Getting it accepted was not easy. The experts that had to judge our article were extremely sceptical. Some of them made it clear that they would never believe in sympatric speciation, whatever arguments we came up with. Others thought that there might still be tiny geographical barriers in the lakes that could have split up the species in an allopatric fashion.'

In the *Biergarten* across the street, a former illegal trades union members' den, Schliewen and Tautz explain that allopatric speciation is extremely unlikely: 'People tend to forget how incredibly small these lakes are: they are round, conical lava pits filled with water', Schliewen says. In other words, there cannot possibly be any geographical barriers there, and even if the water level did drop in the past, the lake wouldn't have become divided into several smaller lakes. 'But the ghost of the allopatric past still roams', Tautz adds with a wry smile. Since their 1994 publication, Schliewen has added a third Cameroonian lake, Ejagham, to his list. Again it's the same story, only this time with an extra twist: geologists have recently calculated a minimum age for Ejagham. 'Only 5000 years', says Schliewen, and drains his pint of beer. 'If this is true, it is speciation at lightning speed.'

So far, Schliewen's DNA studies concur with Ejagham's geological age. He has looked at the DNA of all five Ejagham tilapias and they are practically identical, differing at only a few letters of their DNA code. Because the rate at which DNA mutates is roughly known, Schliewen and Tautz have calculated that the last common ancestor of all five cannot have lived any longer than 10 000 years ago.

Ice-Age lakes

Since the success of Schliewen's Cameroon expeditions, Tautz has developed a special interest in the fish of freshwater lakes. For even though the African lakes present staggering fish radiations, a scaled-down version of the same type of speciation appears to go on in temperate zones. 'We now know lots of examples of speciation in so-called post-glacial lakes', says Tautz, referring to the myriad of freshwater lakes left when the glaciers disappeared at the end of the last Ice Age. Such lakes are found scattered all over mountainous areas in Europe and Canada. In many lakes, two or more closely related fish species swim around, which scientists now believe to have evolved since the lakes were formed, often less than 10 000 years ago.

One such post-glacial lake is the Königssee, a 7-kilometre-long strip of alpine water in the south-eastern tip of Germany, near Adolf Hitler's favourite hangout, Berchtesgaden. Another of Tautz's graduate students, Claudia Englbrecht, studies an ugly little fish called the 'European bullhead' there. Bullheads normally feed on mosquito larvae and other small invertebrates near the surface of the water, rarely swimming deeper than a metre. But when in 1984 a group of researchers from the Max-Planck Institute explored the bottom of the 190-metre-deep lake in the *GEO*, a small two-seater submersible, they were surprised to discover bullheads down there too. It was a mystery what they were doing there: there was no light, only sandy gravel, and no food except for the larvae of water fleas. Tautz and Englbrecht decided that this could be similar to the sympatric speciation in the Cameroonian crater lakes. To prove this, they needed to see whether there are any genetic differences between the fish living at the surface and the ones deep down, indicating that they, too, had stopped interbreeding. But how do you catch tiny fish at depths that go far beyond what is possible with scuba-diving?

The scientists soon realized that they would need the help of the team of biologists at the Max-Planck Institute and their submersible, the *JAGO* (the successor to the earlier *GEO*). In its 10-year existence, *JAGO* had been on quite a few breathtaking missions, including capturing coelacanth fish (famous 'living fossils') on film off the coast of the Comores Islands, and charting the bottom of the Mediterranean near Crete. Tautz and Englbrecht persuaded the *JAGO* team to undertake an expedition closer to home, and it was dropped into the Königssee. *JAGO*'s hull is fitted with a robotic arm, and Englbrecht decided that the easiest way to catch the deep-water bullheads, which swim sluggishly around, would be just to grab them with traps made of fence wire. 'We simply placed a trap over every bullhead we saw down there. When we had enough we hoisted them up,' Englbrecht says.

Having collected 35 specimens in this way, and a similar number of surface-dwellers, Englbrecht was ready to start charting their genetic differences. She used a genetic fingerprinting technique and she also looked at a particularly variable bit of DNA in the mitochondria, the cell's powerhouses. Unexpectedly, however, she found little or no genetic difference. This might mean one of two things: either the deep- and shallow-water bullheads belonged to a single population, with individuals shuttling up and down all the time, or the two populations were on the way towards becoming different species. 'I wouldn't be surprised if they simply hadn't separated long enough', says Dolph Schluter, a soft-spoken expert on fish speciation. Schluter, who is at the University of British Columbia in Vancouver, Canada, has been studying coastal lakes similar to the Königssee. After the last Ice Age, with the weight of the glaciers gone, the Canadian land mass rose higher, taking with it parts of some coastal inlets, which became small lakes at a higher altitude. 'Our Canadian lakes are also on the order of 10 000 years old and they are often fairly small. Like they're about 1 kilometre long and 500 metres wide, that's the sort of scale.'

Since the early 1990s, Schluter has been studying three-spine sticklebacks in the Canadian lakes. Sticklebacks are curious fish in that they live both in the sea and in fresh water. It had long been known that the large marine three-spine stickleback often gives rise to so-called 'limnetic' freshwater populations in coastal lakes. These freshwater forms are smaller and many have evolved into species separate from the marine species. Because the evolutionary tree looks like a fat main (marine) trunk, from which numerous smaller lake-twigs sprout, some imaginative scientists have termed what goes on in these sticklebacks 'fuzzy evolution'.

Some of Schluter's Canadian lakes had landed their own branch from the fuzzy tree, housing, as they do, a freshwater three-spine stickleback—or, rather, *two* species of three-spine stickleback, because each of six lakes harbours, next to the limnetic species, a second, 'benthic' one. The limnetic species is definitely the more elegant of the two: it has a small and slender body, tiny mouth, large eyes, and a long snout. The benthic one is the opposite: large, thick-set, broad-mouthed with small eyes and a stumpy snout. As in the bullheads of the Königssee, the limnetics gobble up the mosquito larvae that float at the surface of the open water of the lake, while the benthics prey on small shrimps and other invertebrates that live in the sediments at the bottom and the shores. When Schluter transplanted limnetics in cages to the shore, they didn't eat as much as in the area where they came from, and when he likewise put benthics in open-water conditions, they fared worse as well.

The two species' mating preferences are different, too. Like Schliewen's African fish, sticklebacks like to mate with partners that match themselves in size. Because the limnetics are small and the benthics larger, cross-breeding between the two is rare. Nevertheless, Schluter and his student Todd Hatfield managed to produce hybrids in fish tanks in the laboratory that were intermediate in build between their parents. When they put these hybrids together with 'pure' limnetics and benthics in cages in the lake, it became clear that in the wild hybrids were always at a disadvantage. When accompanied by benthics close to the shore, the hybrids grew slowest, and when accompanied by limnetics in the open water, the same happened. So wherever in the lake you are, it's always bad to be a hybrid, it seemed. Not surprising, really, because hybrids, being halfway between the benthic and the limnetic species, are not well suited for either environment.

According to Schluter, the lakes offered two open niches to the sticklebacks, and natural selection moulded the fish living and breeding in the open water into limnetics, and the ones by the shore into benthics. The size differences that resulted reduced hybridization because fish teamed up according to size, and any hybrids that still crop up now and again are selected out because they are outcompeted by their better adapted parents.

Speciation seems to have happened fairly quickly, because the benthics occur nowhere else but in these post-glacial lakes. The speciation event that caused them to split off from the limnetics must have taken place within the lakes' lifetime, which is less than 10 000 years. So where did the benthics come from, and how did they manage to spread from one lake to the next? To solve these questions, Schluter's colleague Eric Taylor has been taking a closer look at the DNA of the fishes in several lakes. He extracted rapidly-mutating DNA from the animals' mitochondria and looked for differences between limnetics and benthics from three different Canadian lakes, called Paxton, Priest, and Enos. The result was surprising. Both (benthics and limnetics) are indeed closely related, but limnetics from one lake are more related to benthics from that same lake than to limnetics from another lake, and, vice versa, benthics resemble more the limnetics they share their lake with than they do other benthics.

So this would imply that speciation had happened sympatrically, simultaneously, and independently in all the post-glacial lakes where there are benthics. Schluter has coined the term 'parallel speciation' for this phenomenon. In spite of his discoveries, he is still cautious, because appearances may be deceptive. He is particularly worried about hybrids that may occasionally backcross to the parents, shuttling genes between the two, and making them look more similar. 'We can look at their mitochondrial DNA', Schluter notes, 'and say, aha! either there is gene flow now or it has ceased in the very recent past. And we know that gene flow erases history. So we should at least consider the possibility that it has erased a history that is different from what we think.' So he considers that it might still be possible that the benthic species originated in allopatry (say, in a small shallow lake, where it evolved its adapted body shape) and subsequently invaded a larger lake that had already been populated by the limnetics. 'I think that is completely plausible.'

To distinguish between those two possibilities, Schluter now plans to try to locate the DNA that is responsible for the species' differences in body shape. Those genes will be ones that cannot flow between the limnetics and the benthics because, if they did, they would make each less well adapted. By comparing those genes between the two species and between various lakes, Schluter hopes to be able to give a final verdict. If they are the same in benthics all over, this species cannot have evolved many times independently. Schluter: 'Those are the genes that will tell you the history.'

But even though his own sticklebacks have not fully convinced him yet, he acknowledges that sympatric speciation goes on in many other fishes. In fact, many examples mirror his sticklebacks in more than one way. Everywhere in the Northern Hemisphere, post-glacial lakes are inhabited by species pairs of fish, often from the salmon family, which have divided their lakes in the now-familiar manner: one is benthic, the other limnetic. Schluter keeps a list of examples, which now amounts to over 10 different cases in North America, Europe, and Siberia.

He is most impressed with Eric Taylor's study of sockeye salmon migration in Canadian and Alaskan river systems. 'They have spawned these small, lake-resident forms called kokanee salmon, which use the same breeding grounds as the sockeye', he says. Like cichlids and sticklebacks, salmons mate according to size. And because the sockeyes are a lot larger than the kokanees, most mate within their own species. But some males of both sockeye and kokanee have the habit of behaving in a sneaky way: they will stalk a mating pair of fish and, at the exact moment the female releases her eggs for her mate to fertilize them, rush up and dump their own sperm over them instead. 'So that means there is a certain fraction of eggs every generation that is fertilized by the wrong type. Yet they maintain their genetic differences', Schluter says. This suggests that, like in his sticklebacks, natural selection continuously weeds away the maladapted hybrids. The reason that Schluter does not keep the allopatric option open in this case is that kokanees have been seen to evolve when sockeyes were introduced by humans to river systems where they were not native before. 'They have produced in these drainages new, entirely *de novo* freshwater forms!', Schluter laughs with amazement.

CHAPTER EIGHT

Ecotone—Speciation Prone?

Kinks in the Environment Spawn Species

Fish in cold arctic lakes and in muggy tropical ones have spawned new species because these biotopes offered more niches than a single fish could fill. Monophagous insects have done the same with the abundance of different host plants on offer. This is sympatric speciation in its purest form: first there was a homogeneous gene pool spread over a varied habitat, then the gene pool started to curdle and got lumpy.

But what if, say, apple trees in the Hudson Valley were not randomly strewn in between hawthorn trees? What if there were two separate, abutting fields of apple and hawthorn? What if there had been one gigantic apple orchard bordered by immense hawthorn thickets? Would the same have happened? Would the apple species of *Rhagoletis* still have formed? Yes, it probably would. Although the situation may be anything but identical on a man-made vegetation map, from the fly perspective there would not be all that much difference. Rather than an ecological borderline around each individual apple tree, there would be a single, long and straight transition zone between the hawthorn area and the apple orchard. And the apple flies would still have evolved sympatrically wherever they encountered the border.

Such borderlines—sudden changes in habitat—are called ecotones. For example, if you hike from the town of Arles in the south of France down to the sea, a distance of less than 30 kilometres, the vegetation changes abruptly from Mediterranean maquis to salt marsh. That is an ecotone. Or think of a tropical mountain like Kilimanjaro, which has all imaginable vegetation types crammed together along its 6-kilometre-high flanks, ranging from tropical forest in the foothills to arctic tundra at the summit. Ecotones can be even more drastic than that. Imagine such ecological shocks as where fresh river water suddenly plunges into the sea, or at the entrance of a tropical cave, where a bright, hot, and very rich habitat changes into a dark, cool, and poor one over a distance of only a few hundred metres. Or the chemical disturbance caused by a block of serpentine rocks (poor in nutrients, calcium, and water, rich in nickel, chromium, and cobalt) intruded into a limestone area (which is rich in nutrients, calcium, and water, and poor in heavy metals).

Such sharp ecological breakpoints captured the imagination of Julian Huxley, one of the architects of the Modern Synthesis (see Chapter 3). In 1939, Huxley innocently wrote that if species were living in an area through which ran an ecotone, the populations on either side would evolve lots of adaptations to the different habitats. The intermediates, he wrote, will be 'less well adapted and harmonious', so they 'will remain restricted to a narrow zone, and will not spread progressively through the population'. In Huxley's view, this was as good as speciation. Ernst Mayr, who disliked any sort of speciation theory that was not allopatric, categorically rejected the idea of his fellow Modern Synthesizer. By 1963, he was so convinced that his peripheral isolates model was the only way new species could evolve, that he felt confident to state: 'there is no evidence for, or indeed likelihood of [Huxley's model]'.

But at about the same time that Guy Bush was starting to attack Mayr's allopatric view with the aid of the apple maggot fly, *Rhagoletis*, University of Edinburgh graduate student John Endler was demonstrating that ecotones could, in fact, make sympatric speciation happen as well. Unlike Huxley, Endler, who had a knack for algebra, took a mathematical approach. In 1977, he published a small book called *Geographic Variation,*

Speciation, and Clines filled to the brim with formulas. Where Huxley had used plain but not always convincing language, Endler would say things like, 'A gene *D* causing assortative mating with respect to locus *A* will spread if $[W_1 + W_3 - 2W_2] + (1 - s^2)[a + c - 2b] + (1 - u^2)[d + f - 2e] > 0$'. Even Mayr would have been baffled by that.

Although they used different languages, basically Huxley and Endler were telling the same story. It was an ecotone version of what Guy Bush, Bill Rice, and all the others had been saying about food specialization. Even if the animals were flying, running, crawling, swimming, or slithering across the ecotone all the time, natural selection could be so different in the two habitats, and it could work on so many genes at the same time, that the two populations could start changing. Evolution would be too fast and too pervasive for the migrants to be able to water it down. And along with all the evolution to adapt to the vastly different niches, pleiotropy could kick in sooner or later to change sexual behaviour. Remember pleiotropy (Chapter 7)? Change a clock-gene to suit the life cycle of your food, and—oops!—the time of day you copulate shifts as well. Or grow bigger to catch larger prey—darn!—also just lost your sex-appeal to smaller mates.

Ideas like these have been around for 60 years or more, and even Endler's publications date back to a previous generation. With ecotones criss-crossing landscapes everywhere and at all scales, there should have been ample opportunity to go out in nature and see whether sympatric speciation is going on across them. But, once again, progress in our understanding of speciation has been held up for decades because of the Mayrian fixation on geographical isolation. Tom Smith, an ecologist at San Francisco State University, says: 'It seems so simple . . . Why isn't it something that we've looked at in greater detail before?' Smith knows what he is talking about, because he has been carrying out some of the most convincing ecotone studies to date.

Birds of a feather won't stick together

On the face of it, ornithologist Tom Smith was perhaps not the most likely person to start getting into the murky and shark-infested waters of speciation debates. But on second sight, his knack for ecology probably made him the best man for the job. In the 1980s, Smith was doing his Ph.D. on finches in Africa, especially the feeding and nesting of so-called estrildid finches—relatives of those zebra finches that Nancy Burley dresses up with paper hats (Chapter 4). 'My training was mostly in ecology', says Smith, who now heads the Center for Tropical Biology at San Francisco State University. One of the areas where Smith found himself doing field work was central Cameroon. Here, the wet, tropical rainforests that clothe the country's coastal regions gradually change into the dry savannah of the north. The ecotone between these two habitats is a patchwork of bits of dry forest and stretches of grassland. 'The forest patches in these areas almost represent islands, floating in a sea of savannah', says Smith.

While he was studying finches there, Smith began to realize that these patchworks might be interesting to study from the viewpoint of genetics. His initial angle was one of allopatry. 'I immediately thought, well, of isolation and drift. Sort of an island archipelago scenario', he says. So he set out to see if the birds living in the bits of forest in the ecotone, surrounded by savannah, had—due to genetic drift and the founder effect—become genetically different from the ones in the full-blown rainforest down south. To test this, Smith chose a common forest bird, the little greenbul, *Andropadus virens*. His plan was to visit six sites in the ecotone and six in the rainforest, trap greenbuls in mist nets, draw a few drops of blood, and then analyse these with a DNA-fingerprinting technique.

'So I began working up there and sampling these islands and I wasn't paying attention to morphology', Smith recalls. But as he went from field site to field site, he began noticing small differences in the way the birds looked. It seemed that the birds from the rainforest were a bit smaller and stockier than the ones in the savannah. 'These were things that I initially picked up just looking at the individuals', he says. When Smith realized this he went back to all his field sites to do some serious measurements. 'I found the differences were every bit as big as I thought they were', he says. To be precise, his calipers told him that the wings and legs of the savannah birds were on average about 10% longer than those in the rainforest and were 10% heavier too, whilst their beaks were a tiny bit deeper. This may not seem like much, but it is a larger difference than is sometimes seen between different species living in the same habitat. And Smith knew from his previous work with finches that all these traits are highly heritable, which means they are not easily influenced by their food or some other factor from the environment. They are nature rather than nurture. In other words, they must have evolved.

At the same time, Smith's fingerprinting tests showed that the savannah populations were not so isolated after all. Most of the 12 populations proved very similar genetically. From the distributions of DNA profiles, Smith and his colleagues calculated that in every bird generation up to 10 adult birds migrated between the savannah and the forest. So things were precisely the other way around from what Smith had initially expected. 'I started my work pretty much just following the line of, well, it must be isolation here that's important', he says. But his experiments were telling him that his birds *were* evolving, not because they were isolated but *in spite of* the fact that they were exchanging genes. Only when the gene flow was of the order of eight migrants per generation or more, did the morphological difference start to peter out. Smith, who published his results in *Science* in 1997, suspects that the larger wings in the savannah birds are a response to raptors. In such an open environment, the birds are easily spotted by a patrolling hawk or eagle so they need better aerodynamics for escape. Since his greenbul work, Smith says he has found the same pattern in West African finches and sunbirds.

The Cameroonian ecotone is at places more than 1000 kilometres wide—not really what you would call an abrupt transition. But Smith's latest project has been an application of his tried and tested greenbul method to glossy Australian lizards living in an ecotone of an entirely different order of magnitude.

The Pleistocene and the plasticine

Although to many, the Australian landscape may be synonymous with deserts, droughts, and dingoes, the high north of the continent is actually tropical and humid. Along the east coast of the Cape York Peninsula, between Cooktown and Townsville, lies a strip of 7500 square kilometres of full-blown, tropical rainforest. To the east, rainfall drastically decreases. So drastically, in fact, that the ecotone between jungle and the adjacent eucalypt forest is often as narrow as only a few hundred metres.

In 1997, Chris Schneider and Craig Moritz of the University of Queensland in Brisbane teamed up with Smith to try his greenbul method on the *Carlia rubrigularis* lizards they were studying up in the Cape York area. These reptiles, also known as four-fingered skinks, dwell in leaf litter of both rainforest and eucalypt woods. They live on either side of the so-called Black Mountain Corridor, a geographical barrier that cuts right through the area. The barrier, says Schneider, is known to have been in place since at least the start of the Pleistocene epoch, around 1.6 million years ago. So Schneider located eight field sites: four on each side of the Black Mountain Corridor, two of them in rainforest, the other two in eucalypt forest. He then went ahead and did basically the same that Tom Smith had done with his greenbuls: trapping mature animals, taking tiny tissue samples for DNA analysis, measuring their body proportions, and then releasing them again. When Schneider analysed his data and discussed them with Smith, the latter must have experienced a sense of *déjà vu*. Just like his birds, the skinks' DNA profiles showed that animals were migrating across the ecotone all the time. And yet, natural selection had moulded their appearance considerably—the ones from the eucalypt forest were generally smaller, with relatively shorter limbs and larger heads for their size.

There were some added bonuses to the skink story. First of all, the ecotone was about a thousand times as narrow as in Cameroon, and the change in the reptiles' appearance could actually be noticed while strolling through the ecotone. Second, the ecotone model could be tested directly against the geographical isolation model, because of the presence of the Black Mountain Corridor. As it happened, the DNA profiles of skinks on either side of the barrier were very different, as expected for a 1.6-million-year-old obstacle. But in spite of that, there was no consistent difference in shape between animals from the north or the south side of the Black Mountain Corridor. Instead, the same change took place across the ecotone in the north and in the south independently.

A third fascinating result from the skink study was that Schneider did some clever experiments to see what was actually driving the evolution of differences across the ecotone. Suspecting an influence from the abundance of predatory birds in the eucalypt forest, Schneider had his assistants Maurizio and Ida Bigazzi make almost 500 plasticine skinks, which they meticulously painted to make them look exactly like the real thing. Schneider then placed the mock-up reptiles at regular intervals along two lines crossing the ecotone. Twice a week he walked the lines to inspect his dummies. Sure enough, he found 25 plasticine skinks with bill-marks, showing they had been attacked by birds. And almost all of those were in the eucalypt zone. So life out there is harder than in the rainforest, says Schneider, which would have promoted the evolution of maturation at a smaller size. Clearly, in the eucalypt forest, skinks that managed to reproduce at a younger age must have been at an advantage. So although he did not have an explanation for the differences in head and limb size, his experiments did fit well with the fact that the eucalypt skinks were smaller.

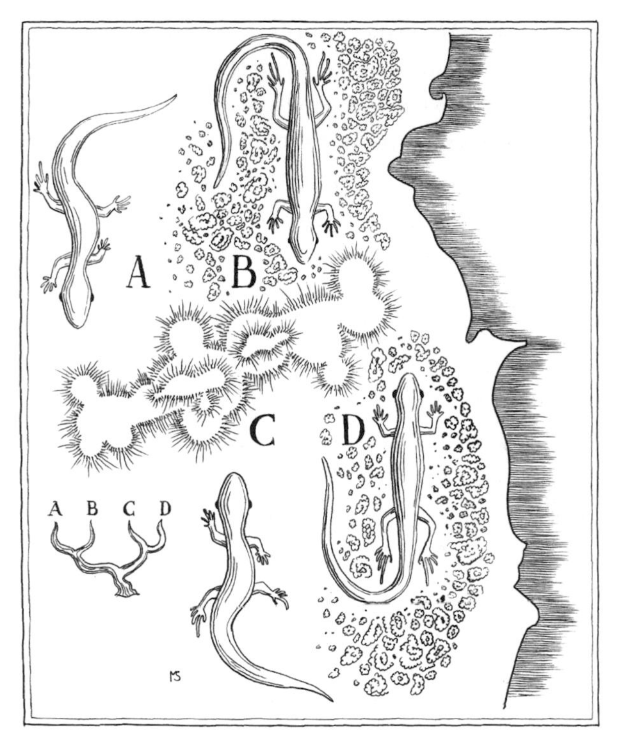

A schematic representation of the differences in skinks at Australia's Cape York peninsula. Skinks on either side of a mountain barrier are only distantly related, and, yet, they show the same differences across the ecotone: those in rainforest are larger and more slender, while those in open forest are stockier.

Studies such as these of skinks and greenbuls show that differences on either side of an ecotone can build up by natural selection, and in the face of gene flow at that. Adaptation to the requirements of the habitats produces differences in the way the animals look and even in their breeding times. At first sight, these results may not be something to write home about. After all, is it really surprising that animals adapt to their local circumstances? Isn't this a basic prediction of evolutionary theory? Yes it is, but many evolutionary biologists had written that this could happen only when there was no gene flow to spoil things. Now it appeared that greenbuls and skinks and who knows how many other organisms had not been reading the literature.

But adaptation is not necessarily speciation. How about reproductive isolation? Is there any evidence that different sexual habits or mating signals would get established on either side of an ecotone and produce reproductive isolation? And, if so, how exactly would this happen? For the answer, we need to turn to plants that thrive on poison.

Heavy metal and hard rock

If you think that toxic waste is something of recent times, think again. For as long as people have needed metals for their axes and daggers and jewellery, they have been mining for them, polluting the environment as they did so. For example, copper was being mined in the Balkans as long ago as the fourth millennium BC, and the zinc mines in Belgium are known to have been opened by the Romans. Like us today, those erstwhile miners were not really worried about the waste from their activities. Large spoil heaps usually mark the sites where they extracted metal from its ore. As the smelting process was not always very efficient in those days, these mine-tailings are still very high in toxic heavy metals. For example, the area around Devon Great Consols in Britain, the richest copper mine of Europe in the nineteenth century, is heavily contaminated. In places, the copper concentration reaches 1%, and the concentration of arsenic may be higher than 5%.

One might expect that such spots are lunar landscapes, where no plant or animal can survive. But one would be mistaken. There is life there, but not as we know it. Without exception, plants growing on these highly contaminated soils have evolved to become heavy-metal tolerant, says Mark Macnair, a botanist at the University of Exeter in England. For more than 25 years, Macnair has been studying heavy-metal tolerance in plants on spoil heaps of copper mines in Britain and California. Well, not just *on* the spoil heaps, but also around them. After all, the boundary between the normal soil and the toxic soil is as severe an ecotone as you can get. And what goes on at this ecotone is fascinating evolution.

In 1983, Macnair discovered that copper tolerance in the yellow monkey flower, *Mimulus guttatus*, is produced by one single gene. Plants growing on the barren dumping grounds of the so-called Copperopolis mine in California all carry an allele at this gene that allows their roots to cope with the poisonous soil. In addition, the Copperopolis plants have some helper-genes, which reinforce the effect of the major gene. Outside of the mine boundary, these genetic protections disappear.

The presence or absence of copper is not the only ecological transition at the ecotone. It brings an entire suite of differences in its wake. Because of the heavy metals, few plants, animals, and microbes can live there, so there is not much plant debris. What little there is, is not broken down, so a decent soil cannot build up. Because of the lack of soil, water and its nutrients and minerals filters straight through, which makes the mines very dry and poor. 'So you get this combination of drought, low mineral status and high metals', says Macnair. 'It's a very harsh environment.' Consequently the toxic desert of the mines poses more than just a single problem for the plants growing there. As Macnair discovered, *Mimulus guttatus* in some Californian copper mines come into flower more than two weeks earlier than the ones outside the mines. In the unbearable summer drought,

Mimulus guttatus and *M. cupriphilus*

these plants have evolved the strategy of producing flowers while they still can.

In fact, drought tolerance normally evolves hand-in-hand with heavy-metal tolerance, says Macnair. And he tells the story of the British National Seed Development Organization, which in the 1970s was developing salt-tolerant grass varieties. These were supposed to be sown along roadside verges that had become so salty from winter-gritting that regular grasses would not grow there any more. To test some varieties, they planted all kinds of salt-tolerant grasses in a salt-marsh and, just for fun, Macnair says, they put in a grass from the Trelogan mine in Wales, which is not salt-tolerant but zinc-tolerant. As it happened, the year was 1976, which had the driest summer in Europe for decades. 'At the end of the summer the whole of Britain was sort of dark brown—with all the grasses dead. The only green spot in the whole of southern Britain were these zinc-tolerant grasses growing on this salt-marsh. Some of these mine plants can retain their green colour through quite severe drought.'

Apparently, then, on either side of the narrow ecotone bordering contaminated soils, the evolutionary climate can be quite different. 'Natural selection there is towards all sorts of things', Macnair says. In fact, on the McNulty mine in California he has discovered a monkey flower that seems to be an entirely new species, an evolutionary offshoot from *Mimulus guttatus*. It is copper tolerant, of course. But unlike *guttatus* it is small and branched and has pale-yellow flowers that have a different shape. 'I went to all the taxonomists and said, what is this? And they said, we haven't the faintest idea; you'd better describe it as a new species.'

The new species, which Macnair named *Mimulus cupriphilus* (the copper-loving monkey flower), does not just look different. Its breeding strategy is unusual as well. For a start, it is an annual: it does not overwinter. Furthermore, it flowers even earlier than copper-tolerant *M. guttatus*, almost a month before the *guttatus* on 'the outside'. Whereas these normal plants finish flowering no earlier than late May, the *cupriphilus* plants are 'sort of dead by the end of April'. And unlike *Mimulus guttatus*, which relies for its fertilization on the local bumble-bees, *M. cupriphilus* has gone self-sufficient. Bypassing any pollinating insects, these plants short-circuit their flowers by letting their own pollen fertilize their own ovules. This is not surprising because bumble-bees simply aren't around at the early date at which this species gets into flower. The upshot is that hybridization between *Mimulus guttatus* and *M. cupriphilus* never occurs.

Because *Mimulus cupriphilus* has only been found on the McNulty mine, and because this mine was opened in the wake of the Gold Rush, in 1849, Macnair suspects that it may have evolved from *M. guttatus* during the past 150 years. The two are very similar in their DNA, so they are definitely closely related. If so, what happened probably was the following. Copper-tolerant *M. guttatus* plants on the mine evolved early flowering. Because their change in schedule meant they were missing the pollinators, they evolved self-pollination. Once this was accomplished, the mine plants could no longer exchange genes with the outside plants any more, after which they could start evolving their smaller size and their change from an overwintering to an annual plant. With the soil so poor and flowering taking place necessarily very early in the season, there was no other option than to live fast and die young: a true heavy metal philosophy.

Back in the field

Meanwhile, tell-tale signs of speciation across ecotones are turning up everywhere in all kinds of obscure creatures. Spanish and Swedish researchers studying snails called periwinkles along the Atlantic coast, for example, are claiming that new species are forming at the interface between wet, wave-exposed rocks, and the ones higher up. And even caves, once considered the classic stage for geographical speciation (see Chapter 2), are starting to be seen in a different light as well. Stewart Peck, a Canadian expert on cave animals at Carleton University in Ottawa, and a self-proclaimed adherent of allopatric speciation, is now beginning to find evidence for sympatric speciation in insects, spiders, and fish in tropical caves.

Peck, who heads a large survey of the insects and other arthropods of the Galápagos archipelago, has discovered at least 10 cases where a cave-inhabiting species has its presumed ancestor still living at the cave entrance. The caves on these volcanic islands, by the way, are very different from the limestone caves in the French Pyrenees. They are so-called lava tubes, formed when the outside of a lava flow cools and forms a crust, while the inside remains liquid and drains out. 'Have you ever been in a lava tube?', Peck asks. 'Well, you should do that sometime. They vary from small things that you could put your arm into to 10, 20 metres in height. They can be very large. The larger ones are more apt to collapse, though', he adds.

Coryssocnemus jarmila is one example of the Galápagos cave fauna that Peck has been studying. This spider makes its webs only in a single lava tube on the island of Santa Cruz. It has all the hallmarks of an advanced stage of cave evolution: no eyes, less pigment, and skinny legs. In the broad daylight outside the lava tube lives *Coryssocnemus conica*. It has none of these cave features, but otherwise is identical to its underground relative, even up to its genitalia (which, in spiders as well as in other insects, usually carry the most conspicuous species characteristics—see Chapter 4). 'These things have been collected by people within a few tens of metres of each other', Peck says. He thinks that it is very likely that the outside species gave rise to the cave species due to adaptation across the ecotone. But, so far, he has not been able to prove this genetically, because the cave species is quite rare. 'It's the kind of situation that to get a good sample it would be necessary to go in and clean out the cave'.

But other tropical cave situations look more promising. In Mexico, says Peck, two closely related fish live in a stream system that runs through a cave. Inside, the fish are of the familiar pale and blind type, while outside they are normal-looking fish. 'At certain flood seasons', says Peck, 'the outside fishes can swim upstream into the caves and the cave fishes get washed out. There is the possibility for gene flow to occur occasionally. But it seems that the outside and the cave populations have such extreme selective pressures.'

Naturally, John Endler (now at the University of California in Santa Barbara) is pleased that, after almost 25 years, his theory is finally catching on. He blames the dogma of geographical speciation for holding up progress. 'Now that [those] strong opinions have basically disappeared and no longer dominate the field, people are starting to take other ideas seriously', he says. Endler also thinks that the lack of field biology has been frustrating. 'Anyone who works on real animals up in the real world can see these incredible effects of the environment all the time', he complains. 'It's hard to escape it. But people in the lab don't really believe the environment is important.' Taxonomists, too, should not just study the animals in the museum, he says, but 'actually watch them, look at their behaviour, and think about how they are affected by the environment.'

Once again, it appears that adaptation to the environment is all-important. So let us take stock here. We have seen in insects and fish that speciation can happen in sympatry, when a fraction of the population settles into a new niche and natural selection causes a schism. In this last chapter, it has turned out that the same can go on across an ecotone in things as diverse as lizards, birds, and plants. In allopatry, on the other hand, sexual selection seems to be a key ingredient, as it drives isolated populations in different directions (Chapter 4). In Chapter 10, we shall see that ecology is important in allopatric speciation as well, but what about sexual selection in sympatry? The subject has been somewhat absent from our explorations of sympatric speciation. In the following chapter, however, this will be remedied.

CHAPTER NINE

Victoria's Blue Genes

Sex in Sympatry

In the early nineteenth century, King Mwangonde walked from his home town to the village of Mwela to marry Mapunda. To do this, he crossed a plain that today lies submerged beneath the clear waters of Lake Malawi, the southernmost of the three large East African lakes, dubbed the Great Lakes. In 1894, a traveller named Swann saw gigantic trees growing in the water near the shores of the lake. And in 1983, the centenarian chief Moses Ngosi told a British diplomat that his grandparents used to talk about a sunken village that lay an hour's walk lakewards from the present-day Ruasho River delta. Historical records like these confirm what geologists now know for a fact: the water level of Lake Malawi has risen by some 120 metres in the past two centuries.

When Mwagonde went on his marital quest, he probably passed a cluster of rocky pinnacles that have since become offshore islets. Had he been a clairvoyant, he would have been able to see the spectres of future fishes darting among the grey boulders towering high above his regal head. In the 1980s, long after Mwangonde and his empire had been forgotten, a group of geologists and biologists returned to this spot to study the lake, its geology, and these rock-dwelling cichlids, or *mbuna*, as the local fishermen call them.

There are some two hundred different *mbuna*, comprising 40% of the total cichlid fauna of the lake. *Mbuna* are poor travellers: usually they do not stray from the rocky patch where they are born. Although it had been known that all these cichlid species are unique to Lake Malawi, the researchers discovered that they are endemic even to the single rocky island they inhabit: Likoma Island, halfway along the eastern coast of the lake, for example, has 39 different species swimming around its shores, 25 of which occur nowhere else in the lake but there. Yet Likoma, and many other islands like it, were dry piles of rubble little more than a century ago. The inescapable conclusion is that their endemic fish faunas developed in the short time-frame since the shore of the expanding lake first reached the foot of the rocky hills.

Now the question is, did the species that now populate these isolated rocky islets speciate in sympatry, dividing up the ecological niches like an inverted Cameroonian crater-lake scenario in a startlingly short time, or did they evolve somewhere else, colonize the islands independently, and die out elsewhere? The way to test this would be by looking at the genetic relatedness of the endemics around the islands. If they split from a common stock less than 200 years ago, their genetic differences should be minimal; if they represent independent colonizations, they should be unrelated.

Peter Reinthal, a biologist at the American Museum of Natural History in New York, did just such a test. He collected fish around Likoma Island and at Otter Point, a rocky outcrop at the southern edge of the lake. At Likoma Island, he focused on four *mbuna* species that had distinct ecological roles. Because the fish were new to science, they had no official names yet, so the researcher gave them nicknames. All ate algae, but 'Red Cheek' foraged in shallow water without sediment on rocky surfaces; 'Membe' preferred deep water, rich in sediment, on a sandy bottom; 'Yellow Chin' went for the same habitat as Membe, but in shallower water; and 'Dark,' finally, hung out in deep water, over rocky surfaces with not too much sediment.

At Otter Point, Reinthal identified four species of *mbuna* that were very similar to the quartet he had singled out at Likoma.

For each of the Likoma species, he could point at an Otter Point species that matched it both in appearance and in ecological preference. So it seemed very likely that each ecological duo had a common ancestor. Alternatively, each quartet might have had a common ancestor, unrelated to each other, and their similarities were due to parallel evolution. To solve the puzzle, Reinthal teamed up with molecular biologist Axel Meyer (then at the University of California at Berkeley), who had been successful in the past at trying his molecular hand to solve problems of cichlid evolution.

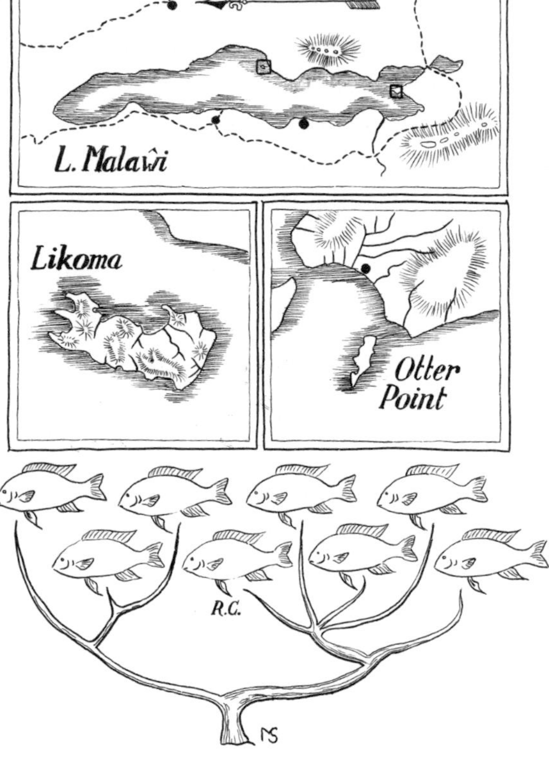

Rock-chichlids in Lake Malawi. The species at Likoma Island are closely related, and so are those at Otter Point, with the exception of 'Red Cheek' (RC).

Together they selected a fast-mutating chunk of DNA from the fishes' mitochondria, which should show differences even if the species were closely related. They determined the DNA sequences from three individuals of each of the eight species and then produced an evolutionary tree. As usual, the result was not as clear-cut as the researchers might have hoped, but a few things stood out: three of the Otter Point species and one of the Likoma species (Red Cheek) were genetically identical. The fourth Otter Point species was a little different, but also quite similar to these four, while the remainder of the Likoma Island species formed a trio by themselves. According to Meyer, 'This shows that, possibly with the exception of Red Cheek, each island assemblage evolved at their present-day locality in a very brief time-period.'

At first glance, it would seem that speciation around each of these rocky islets proceeds much the way it has in the Cameroonian crater lakes, with competition for food forcing a population apart in characteristics that are also important in mate choice. But there are some important differences. In the Cameroonian cichlids, size is a major factor: the largest tilapia in Lake Bermin is almost 20 times as massive as the smallest one. Also, most of them have more or less identical coloration, and males and females form long-lasting mating pairs. The rock-dwelling cichlids from the great African lakes, on the other hand, stand out because of their vivid and strikingly different colours and their promiscuity.

Small fry

In the Institute of Evolutionary and Ecological Sciences of Leiden University in the Netherlands, a decorative fish tank is quietly bubbling away in a little-frequented corridor. In it, a few Lake Victoria rock-cichlids play hide-and-seek among algae-covered bricks and logs. They are stunningly beautiful: some are checkered black and white, others are as orange as chunks of papaya, while only a velvety black one with blood-red fins seems to honour the name the Kisukuma fishermen have given them: *mbipi*, the dark ones.

Two flights of stairs up, and the same colourful fish are everywhere: on posters on the walls, in jars on the tables, and on the computer screen of graduate student Ole Seehausen. His career is surprisingly similar to that of Uli Schliewen, who studied the Cameroonian cichlids. He, too, is German, with a life-long passion for cichlid fish. And like Schliewen, Seehausen has betrayed his more than professional interest by writing books for aquarium enthusiasts. His latest is called *Lake Victoria Rock Cichlids* and it is the first book entirely devoted to the *mbipi*, Lake Victoria's answer to the Malawi *mbuna*. The main characters in the book bear names that would not be misplaced in a second rate gangster movie. Meet Thick Skin, Duck Snout, and Fleshy Lips. Nicknames like these are necessary because almost all of the fish in Seehausen's book are completely new to science and have not yet been awarded official scientific names.

Until recently, the *mbipi* were thought not to exist. Humphry Greenwood, the Natural History Museum's late fish expert who discovered more than 100 Lake Victoria cichlids in the 1950s and 1960s (see Chapter 2), found only a single rock-dwelling cichlid. That is why he was convinced that fish like the Malawi *mbuna* were absent from Lake Victoria. So it came as a surprise to Greenwood when, in the late 1970s, the group of fish biologists from Leiden University found a dozen such species in a single bay of Lake Victoria. His surprise continued: between 1986 and 1991, the Leiden group discovered a further 10 species; and since 1991, when Seehausen and his colleague Niels Bouton started their work, the stream of new *mbipi* flowing into the Leiden laboratory acquired torrential proportions. Seehausen's book contains 173 species and he predicts that many more will follow. An entire species flock had been overlooked for decades.

Lucky for Seehausen, most of his darlings have managed to survive the onslaught of the Nile Perch, a gigantic 'vacuum cleaner' fish that was introduced to Lake Victoria in the 1950s for fisheries purposes but proceeded to gobble up all the native cichlids. These predators prefer the open waters of the lake, and although Seehausen has regularly seen them patrolling the rocks where his *mbipi* live, they seem to have made less of an impact there. So he can still study his fish in all their glory. And glorious they are. Mesmerized by the underwater rainbow that unfolded before his eyes and in his nets, Seehausen began to notice something peculiar. Quite often he found, at a single rocky shore, fish that were very similar apart from their colour. For example, *Lithochromis rubripinnis* lives in an area of the lake named Mwanza Gulf. Seehausen describes the hue of the males dramatically as 'bright metallic iridescent sky blue, with bright blood-red fins'. In exactly the same area another fish lives, called *Lithochromis rufus*, of which the males are 'bright blood-red' with yellowish to greenish markings. As in all Great Lake cichlids, the females of both species are a dull brown.

Even though they live in exactly the same places and look like differently coloured copies of each other, Seehausen found tiny shifts in their ecology. *Lithochromis rufus* lives close to the shore, never deeper than 4 metres, where it eats caddis fly larvae and small snails; *L. rubripinnis* is a little more adventurous, daring to go into water 8 metres deep, and avoiding the shoreline. Its menu is concomitantly different also: *L. rubripinnis* skips the snails. The *rufus/rubripinnis* case is not unique. Time and time again, Seehausen discovered species pairs with similar body forms that occurred together, did slightly different things for a living, and had male coloration that separated them into a red and a blue species. In total, he discovered seven such duos, and other cichlid workers have found eight more examples in other habitats in Lake Victoria.

Surprisingly, though, the same division in red and blue also occurs within, rather than between species. Seehausen found no less than 11 species where, in the same locality, there were blue males and red males. But how could he be certain that these were colour forms of the same species and not two very similar 'sibling' species? 'This is indeed one of the questions I am asked frequently', he says. 'First of all, colour is the only thing they differ in. Their teeth, scales, shape, and stripe-patterns are identical. Second, if you take them to the lab they often do not breed true: a single female may throw up offspring of both colour forms.'

The species pump

The parallel that Seehausen saw was intriguing: could it be that pairs of different species started out as a simple sympatric colour variability within a species? And if so, how did it work? First of all, he realized the importance of the breeding system of the Lake Victoria cichlids. Like some of the tilapias in Cameroon, these cichlids are mouthbrooders. But unlike their West African brethren, both sexes of East African lake cichlids are decidedly wanton. Males will share their sperm with as many females as they can find, and females often carry clutches of half-brothers and half-sisters in their mouths; all her own children, but sired by different males.

In such a situation, as we have seen in Chapter 4, sexual selection becomes important. In a monogamous tilapia from Lake Bermin, individuals will not easily gain the edge by being more attractive: these fish form long-lasting mating pairs, so there's a Jack for every Jill. But in the Great Lakes in eastern Africa, things were different. An attractive male will be at an advantage, because he will be able to persuade many females to use his sperm. First, Seehausen had to prove that colour was indeed an important cue for females to select her mates. It had already been known for some time that *mbipi* and related cichlids have a dual sensitivity in the retina of their eyes: red light and blue light are picked up best. So blue and red males might indeed hit females' sensory biases (see Chapter 4). But experimental evidence was also needed. So Seehausen selected a common (and very pretty) cichlid, *Pundamilia nyererei* (yes, named in honour of the former Tanzanian president Julius Nyerere), of which the males come in a red and a blue form. Females are always a drab, brownish grey.

'It has been a big experiment!', Seehausen sighs with the fulfilment of scientific labour. He collected females and males from both colour forms around the small Luanso Island in the Mwanza Gulf and used them in an aquarium experiment in the lab.

Sixteen females were put in tanks where they could choose between a red and a blue male to mate with. Seehausen observed which choices the females made, and, to make sure that the females did not act on a whim, he repeated the tests several times with each female. 'It is great stuff', he says. 'Most females significantly preferred blue males, some were inconsistent, and some preferred red males.'

Apparently, then, some females are partial to a particular coloration in their sex partners. It is a Fisherian fashion, as we have seen previously. This makes it extremely easy for colour variation in the males to gain a foothold. Imagine a cichlid population in which there are only red males, even though some females would prefer them blue. As soon as a mutant blue male appears, it will be extremely lucky in love. It will be the favourite mate for all the females that previously could only dream of such a sexy male, and it will be able to father many children. As a result, in the next generation quite a few male fish will be sporting blue genes.

The same sorts of mate preferences also play a role in keeping species apart, Seehausen found out. He took fish from two different species, one with red males, the other with blue males. On either end of a 3-metre-long fish tank, he arranged small transparent enclosures, which each held a single male: a red male on one end, a blue one on the other. In the centre, a female was released, belonging to one of the two species, and her actions were filmed with a video camera. To avoid disturbance of the fish, Seehausen watched what happened on a video screen from behind a black curtain. When the aquarium was illuminated with white light, the two species bred true: the females of the red species only showed interest in the red males, and vice versa.

But when Seehausen put orange colour filters in front of the light tubes over the aquarium, the females got confused. Because the entire aquarium was submerged in orange colour, like a street lighted with sodium lamps, blue and red both looked like shades of grey. In other words, the colour difference between the red and blue males all but disappeared to the human eye. And to the fish eye as well: none of the females managed to pick out the males from their own species any more, and were just as likely to team up with a blue as with a red male. So colour may be the only barrier that prevents closely related cichlids from hybridizing.

How, though, could a population of fish of the same species, in which different colour forms exist, split into two separate species, each with a different ecology and a different male coloration? In Chapter 4, we saw that the most sexy males will normally take over the whole population. But here, this did not seem to have happened.

In Britain, another cichlid enthusiast, George Turner of the University of Southampton, has been letting his computer crank out simulations of such sympatric speciation by sexual selection. As it turned out, sexual preferences could result in separate species, provided that there were not too many different genes involved in both coloration and female preference. This way, the offspring of females that prefer, say, blue males, will inherit both the genes for blue colour (from their father) and the genes for blue preference (from their mother). As a result, they will start mating among themselves, and will become genetically isolated.

Seehausen is now doing crosses in the lab to work out how coloration and sexual preferences are inherited. Things are turning out to be more complicated than anticipated, with both female *and* male colour preference (in some species, females can differ in coloration also), and a predator kicking in (some colour forms are so conspicuous that cormorants could systematically make a lunch out of them). Nevertheless, it is clear, says Seehausen, that the genetic basis of coloration is simple, with only three genes involved. He also has evidence that mate choice is on the way towards splitting up a population of *Neochromis omnicaeruleus*, in which many different colour forms coexist, some of which seem to reproduce more or less independently of others. 'The way I see it', says Seehausen, 'sexual selection in cichlids is continuously splitting up populations. It acts as a species pump: new species are being churned out all the time.' Many of the newly generated species will go extinct again. After all, differing only in coloration, they will have to compete with the resident colour form. But sometimes, slight ecological differences will evolve, which can help the new species make it on its own. For example, red light penetrates deeper into the water of Lake Victoria than does blue light. A genetic penchant for prey that live in deeper water may spread in the red species. As a result, the red and the blue species will become adapted to slightly different niches, and may continue to exist side by side.

Seehausen's sympatric species pump is still somewhat controversial. Some experts think that the Victoria-cichlids are more likely to have been driven into their respective niches by competition for food, just like Schliewen's Cameroonian cichlids or the sticklebacks. Two back-to-back papers in the 22 July 1999 issue of *Nature* show that virtual cichlids in computer simulations behave this way. First, the cyberfish adapt to different food sources, and only then does sexual selection kick in.

So the jury is still out on Seehausen's hypothesis. Was food competition or mate competition the driving force behind the evolution of Lake Victoria's 500-plus cichlid species? We cannot be certain yet. One thing is clear though: it all must have happened sympatrically, within the confines of the lake. The alternative would be Humphry Greenwood's allopatric Lake Nabugabo scenario (see Chapter 2). For this to have happened, lots of satellite-lakes should have formed all around Lake Victoria's shores. Each of the satellites should have been isolated for a couple of thousand years while new species evolved in them, after which a rise in water level reconnected them with the big lake so the freshly cooked-up species could be released. And this should have happened over and over and over again.

The scenario was hard to imagine in Greenwood's days, when Lake Victoria was thought to be half a million years old. Would sufficient fluctuations in lake level have taken place in such a short period for all the species to have formed? But in 1996, geologist Thomas Johnson and his team definitively pulled the plug on the hypothesis. They drilled into the bottom of the lake at several places and discovered that wherever they drilled, they brought up soil that could have formed only when the lake was dry. The most recent of these soils was only 12 400 years old. In other words, Greenwood's allopatric speciation scenario for the most impressive species flock of the world had simply run out of time.

This brings us at the end of our grand tour of speciation. We have covered a lot of ground. We have seen situations ranging from a single apple tree to the Isthmus of Panama. We have explored the evolutionary consequences of competition for sex and food, in organisms as diverse as the spangled drongo and the copper-loving monkey flower. But what does it all mean for the history of life on our planet? How does it all tie together? That is the focus of this book's final chapter.

CHAPTER TEN

Mystery? What Mystery?

Douglas Futuyma, whose 1980 criticisms set off the second and fruitful wave of *Rhagoletis* investigations, as Chapter 6 recounts, once wrote that speciation is 'more thoroughly awash in unfounded and often contradictory speculation than any other single topic in evolutionary theory'. We know a thing or two about that. In this book, we have had our share of contradiction and speculation. Just think of the debate about bottlenecks, the controversy surrounding sympatric speciation, and the speculations about instant speciation. In fact, speciation research has always had a strange and somewhat disturbing obsession with emphasizing and classifying the differences between the various theories, perhaps because the field emerged from taxonomy, the branch of biology whose business it is to name and classify things.

For example, we have already come across allopatric and sympatric speciation. What happens across ecotones has sometimes been termed 'parapatric' speciation, while Ernst Mayr's bottleneck model in peripheral isolates is also known as 'peripatric' or 'micro-allopatric' speciation. Some sorts of allopatric speciation also go by the name of 'dichopatric' speciation, while certain types of parapatric speciation are called 'allo-parapatric' speciation. To make the confusion complete, when large chromosome mutations are involved, some people have used the term 'stasipatric' speciation. Sadly, the something-patric fashion is just the tip of the iceberg. A perusal of the evolutionary literature reveals a cornucopia of terms and theories, including microgeographic, semigeographic, single-gene, selfish-gene, symbiont, centrifugal, reinforcement, vicariant, competitive, dumb-bell, hybrid, ecological, adaptive, and—believe it or not—Adam and Eve speciation.

Confused? Do not despair. Many are only different words for the same thing, and others are just special cases of more general types of speciation. Nevertheless, it seems that speciation is not a single phenomenon. Even in this book, we have come across a variety of ways new species can and do evolve. With or without geographical isolation; with or without sexual selection; instantaneous or gradually . . . No wonder that an article in *Science* magazine covering a 1996 speciation congress carried the telling title, 'On the many origins of species'. But are the differences really so stark? For example, let us have a closer look at that age-old divide in speciation: the one between sympatric and allopatric speciation. What follows is a story that bears all the hallmarks of sympatric speciation. Or does it?

Treehoppers hop trees

Before *Rhagoletis*—the apple maggot fly—became the benchmark for sympatric speciation, a group of curious cicadas held that title. Cicadas are best known for their job in background sound-effects for films set in the tropics, but a family called Membracidae or treehoppers also stand out because of the 'pronotum', the armoured plate that covers the insects' neck. In treehoppers the pronotum is often shaped bizarrely, sporting spikes, bulbs, crests, or lumps, which make some of these animals resemble woody thorns or other outgrowths of trees—a useful sort of camouflage when you are spending most of your time sitting on branches, as treehoppers do.

Tom Wood, a cheerful entomologist from the University of Delaware in Newark, has spent most of his life in the company of treehoppers. He has been especially infatuated with an eastern North American species called *Enchenopa binotata*. This brownish bug, almost one centimetre long, is the proud owner of an impressive forward-leaning pronotal hump, which makes the insect resemble Dopey the Dwarf in one of his less-stable moments. 'It was love at first sight', says Wood, remembering the fall day in 1965 when *Enchenopa* walked into his life and never left. 'I was pinning some bugs, and apparently I had picked up one of these *Enchenopa*s, when the little kid next door came over. He looks at it and says, "what's that?". I say, "that's a treehopper". And he says, "those are all over on the tree at my front yard". So I went over and sure enough there they were, they were all over the place.' Charmed by the odd-looking insects, and fed up with his research in pesticides, Wood decided to do a little side-project on their natural history. 'It just started out of mindless curiosity', he admits.

Like most treehoppers, *Enchenopa binotata* feeds by sticking its syringe-like mouth into the branches of trees and tapping the nutrient-rich sap streams. When Wood started his treehopper project, the insect was considered to have unusually catholic tastes in that it fed on a large array of American trees, including black locust, redbud, hop tree, black walnut, butternut, bittersweet, tulip tree, *Viburnum*, and hickories. But Wood noticed something unusual: 'I started getting out in the field and looking at these things and, Jeez, the positions where females laid their eggs were different on different tree species, and the larvae, which are called nymphs, were differently coloured also.' Wood subsequently discovered as well that there were subtle but consistent protein differences between treehoppers from different trees. And females that were forced to remain on the 'wrong' tree either died or produced hardly any offspring. So all in all, what had seemed like a single treehopper species with just a very broad taste, turned out to be a complex of no less than nine 'sibling species', each specialized for life on a single or just a few host species.

These treehoppers' specializations go further than just food requirements. For example, the eggs of each of the nine members of the *Enchenopa binotata* clan hatch at different times in the spring, triggered by the flowering of their respective trees. So their developments are out of sync from the onset. This mismatch is exaggerated even further when mating time arrives, because some species start mating only 50 days after hatching, whereas others are not interested in sex until they are 80 days old. As a result, each species has its own narrow time-frame in which they mate, so individuals of different species hardly ever come across one another. Even more so because mating in most species takes place at different times during the day, which may have come about in the same way as in the melon flies described in Chapter 7.

Presumably, says Wood, each speciation event leading to a species with a new specialization, is preceded by a treehopper accidentally hopping on to a new type of tree. But, as he found out, treehoppers are hardcore stay-at-homes, usually spending most of their life on a single tree. Migration is not something they like to do, not even between trees standing next to each other. So once such a host-shift is made, the decision is not easily reversed.

To test his idea, Wood took females from each of six hosts and placed them on non-native host trees. In most cases the females refused to lay any eggs, but, sometimes, at least a few eggs were laid. Wood guarded those eggs carefully and, the following spring, was surprised to see what happened when the eggs were due to hatch. 'Their hatching times had shifted in the direction of the native treehopper', he says. 'For example, eggs from *Viburnum* females, laid on redbud, hatched much later than eggs on the regular host. In fact, they hatched at the same time that eggs of the redbud species normally hatch.' Wood later found out that this shift was entirely due to the trees' yearly water-pumping schedule. Redbud starts regrowing its leaves a little bit later in spring than *Viburnum*. The eggs, which the female deposits right on top of the water-carrying vessels in the branches, imbibe the ascending water, and this is the trigger for the young nymph inside to emerge, because it signals that there will be sufficient young leaves for the insect to eat.

Host-shifts like these are probably exceedingly rare, Wood admits. The treehoppers are very particular about which tree they go for and rarely make mistakes. This is leaving aside the unlikely event of a male and a female both making the same mistake at the same time on the same tree. On an adopted host, females don't easily lay eggs and, when they do, the nymphs often don't fare very well. On the other hand, once all these implausible conditions are met, the migrant treehoppers will be isolated from their ancestors for exactly the same reasons: little dispersal, mating on the host, and strong natural selection to adapt. On top of that, the yearly agenda of the new tree will dictate a shift of lifestyle in the migrant treehoppers, which may mean they will never see their relatives on the old tree again, and speciation is well under way.

Living apart together

On the face of it, Wood's nine treehopper species have evolved in exactly the same manner as *Rhagoletis* fruit flies have. An accidental shift to a new host plant, a life cycle that immediately locks into the plant's, followed by many generations of adaptation to a new environment and that is that. But there is a difference. The treehoppers mix even less with the neighbours than the fruit flies do. In fact, once the initial tree hop has been made, the renegade cicadas are completely and utterly and permanently isolated. They sit on trees surrounded by their ancestors, so close they could almost touch them; and yet they are as far from them as two shrimps on either side of the Panama Isthmus. It is the paradox of geographical isolation in sympatry.

Parasites on birds and mammals are another example of such sympatric and yet allopatric speciation. Intestinal worms, feather lice, and skin mites are often specific to one particular species of bird or mammal. New species evolve when a pregnant parasite accidentally finds itself on or in a new host. In essence, it is no different from a bird that colonizes an uninhabited island. What it boils down to is that, gene flow or no gene flow, natural selection will often be strong enough to drive a wedge into the population and split off a new crater-lake fish or fruit fly, or tree-hopper, or parasitic worm, or whatever. Allopatric and sympatric speciation may not be so different after all.

But surely allopatric speciation is different in the sense that it does not *need* natural selection to produce two species? Even without any adaptation, the steady rain of mutation and genetic drift could produce lots of accidental genetic differences between isolated populations. Nancy Knowlton and her snapping shrimps seem a case in point. Like a country-sized scalpel, the Panama Isthmus bisected the populations of many species of shrimps, which, over a period of 3 million years or more, drifted apart to produce Pacific/Atlantic pairs of species. All the same, there may be more to the shrimp story than meets the eye. The sea where the shrimps lived 3 million or so years ago was environmentally more or less homogeneous, but things have changed since then. As they became separated, the two oceans became quite different. The Pacific Ocean is now more variable in temperature and has more nutrients. And the Caribbean is more stable and much poorer in nutrients. That is why there are vast coral reefs in the Caribbean, but not on the Pacific side of the isthmus.

Knowlton also points out that the Pacific has a 6-metre tide, while the Caribbean has hardly any tide at all. Because her shrimps mostly frequent shallow water, the ones on the Pacific side would have had to adapt to the daily stress of being above the water, with the added risk of shorebirds trying to make a lunch out of them. 'There is undoubtedly selection going on', says Knowlton, and she adds: 'There probably are some real physiological differences between these things, which no one has ever studied, to my knowledge.'

In line with this view are experiments with banana flies. For example, the Greek biologist G. Kilias in 1973 collected a few jars full of *Drosophila melanogaster* in the Greek countryside and brought them to his laboratory at the University of Patras. Then he divided the flies over four Plexiglass cages. Two cages he put in a cold, dark, and dry incubator, while the other two went into a warm, bright, and humid incubator. For three years the flies were left in there. They were given fresh food at regular intervals, but otherwise the populations of banana flies lived and bred in splendid isolation for some 80 fly generations. When Kilias finally decided to end the experiment, he found that when he put together males and females from different cages that had been in the same incubator, the flies showed no preference for mates from their own cage—they mated at random. But when he put flies together that came from different incubators, females preferred males from the same environment over those from the other environment. So females adapted to cold/dark/dry fancied cold/dark/dry males and not so much the warm/bright/humid males.

Time and time again, it seems that good old Darwinian natural selection is what speciation is all about, even in allopatric speciation. As Jeffrey Feder, one of the key players in the *Rhagoletis* debate, says: 'Adaptation to the environment is the trigger, shifts are the key; the geographic situation is only icing on the cake.' And he adds: 'I think in some sense the whole argument is overplayed.'

What's sex got to do with it?

That leaves us with the enigmatic sexual isolation. How do we make sense of the evolution of pre-mating and post-mating reproductive failure? The facts are paradoxical. On the one hand, it tends to evolve accidentally, piggybacking on adaptation to the environment. Think of those melon flies whose daily sexual timetables no longer overlapped after they had adapted to longer or shorter life cycles. And banana flies' sexual preferences somehow evolved along with their adaptation to the inside of those

Greek laboratory incubators as well. The large benthic and small limnetic sticklebacks that match their mates for size is another example, as is the shift in flowering times of monkey flowers on spoil heaps. In cases like those, sexual isolation is the direct consequence of adaptation to a different environment.

On the other hand we have sexual isolation that just seems to come out of nowhere. A freak genetic accident in a snail makes it a sexual outcast and sparks the evolution of a new species of mollusc. A goat's-beard plant accidentally doubles its chromosomes and henceforth can only mate with flowers that made the same genetic mistake. A preference for blue males forces apart a population of Lake Victoria *mbipi* fish. These are all speciations that took place purely by sexual isolation, with no obvious link to the environment.

The key word here is 'obvious'. The link to the environment may be less than obvious, but it may still be there. The goat's-beard's polyploids are environmentally different, because their ideal mix of genes of both parents gives them the ability to invade a new and vacant niche. And blue *mbipi* will only evolve in shallow water because blue colour does not penetrate deep in Lake Victoria. Without such niche difference, polyploid goat's-beards and blue fish would have to compete with the residents and probably never gain a foothold.

And remember Fisherian sexual selection (from Chapter 4). Is it really a random and unpredictable process, detached from the world animals live in? Probably not. It, too, has to run the gauntlet of environmental restrictions and obstacles. Male guppies in Trinidad, for example, are drab in rivers where predatory fish abound, but brightly coloured when enemies are scarce. With enemies that are just as keen on bright colours as female guppies, sexual selection can take off only when the male guppies are relatively safe. Birds that live in the tropics, to mention another example, suffer more from parasites, so female tropical birds may be more likely to pick a male for good-gene reasons than for purely aesthetic ones. And cricket frogs that live in dense pinewood in Texas call at a higher pitch than relatives that live in the neighbouring plains, because the trees filter out lower sounds. These are just a few of the many environmental factors that could subtly influence the outcome of sexual selection. So sex and the environment interact.

The interaction between sexual and natural selection might also explain what caused the birth of Mayr's theory of bottleneck speciation. 'My theory was based on observation', says Mayr. 'The empirical fact that the most distinct populations we always find peripherally isolated is very much evidence in favour of my ideas.' But there are other explanations for this than the founder effect.

When Mayr described his bottleneck model, one of his prime examples was, if you recall, the paradise kingfisher (see Chapter 2). This is a very weird kingfisher, with a tail almost twice as long as its body. On the New Guinean mainland, the bird looks the same everywhere. But on each of the small offshore islands, it is different in tail length, in the shape of the tip of the tail, and in the colour of the head and back. Mayr could not explain this from an ecological perspective. The landscape of New Guinea is quite varied, he argued, whereas the forests on the offshore islands of Numfor and Tagula look just the same. So why would the peripheral populations of the paradise kingfisher be so different while those on the vast and variable mainland are so similar? Arne Mooers, the Canadian biologist who did those massive bottleneck experiments with banana flies, thinks there may be more to island ecology than meets the eye. 'We might not see any differences, but the differences might still be there', he says.

> We all know that small islands offer a very different selection than large islands. The wind, the temperature, the salt concentration, the amount of food you get . . . The type of vegetation, the elevation . . . all these things. We don't know what other species made it over to those islands. We don't know what diseases there are that aren't on the mainland . . . Selection is incredibly subtle.

What we end up with is a simple and abstract image of speciation. The process will get started only if the balance between natural selection and gene flow is right. If there is a lot of gene flow (sympatric speciation), you will need strong natural selection. If there is little gene flow (allopatric speciation), natural selection need not be so severe either (although it won't harm if it is). Once under way, speciation will amplify itself with the help of sexual selection. Set off and channelled by small differences in the environment, sexual signals in both populations will start to diverge. As a result, males from the one population will get fewer sexual encounters with females from the other, so gene flow is reduced as well. This will cause adaptation to work even stronger, and so on and so forth until Darwin's mystery of mysteries has unfolded itself once again.

Overrated barriers

But geographical isolation has a final trump card to play: it may have much more opportunity to work. As our abstract of speciation reveals, the easiest way to get speciation going is still by complete geographical isolation, because it needs only slight differences in environment on either side of the barrier to get evolution going. How often will this be so? In other words, what has been the relative contribution of geographical isolation in generating the world's biodiversity?

More than 99% of speciation is allopatric speciation, said Ernst Mayr when he was interviewed by *Science* magazine in 1996. In vertebrate animals it is 94%, claimed John Lynch of the University of Nebraska in 1989, after looking carefully at distribution maps of frogs, fishes, and birds. Michael Rosenzweig of the Universty of Arizona in Tucson also backs geography when he says that it 'usually dominates', while sympatry zealot Guy Bush, understandably, thinks it plays only a minor part. In the sea, where organisms, their eggs, and their larvae float about unhampered, it is also hard to imagine a leading role for geographical isolation. Even Nancy Knowlton admits: 'a big barrier like the Isthmus of Panama makes a trivial contribution to the generation of [marine] species', but she adds that there may be all kinds of underwater barriers that marine biologists haven't fathomed yet.

As usual, then, unanimity is hard to find. But scientific data are starting to support the notion that the impact of geographical barriers has been trivial rather than paramount. Take, for example, the Ice Ages. If allopatry were important, the last Ice Ages should have left a clear mark on the biodiversity of the Northern Hemisphere. And, in fact, many biologists have implicated the huge ice flows of the past few hundred thousand years in producing many of the species of Eurasia and North America. As Michael Rosenzweig, an ecologist from the University of Arizona in Tucson, points out, 'what better barrier than a continent-sized glacier knifing down . . . and slicing many of its major environments into . . . segments?'

But glaciation is beginning to look more and more innocent, as fossil insects reveal. Palaeontologist Russell Coope of the University of Birmingham has built his career on picking fragments of beetles out of peat bogs. Beetles make ideal fossils: their hardy armour can survive thousands, even millions of years essentially unscathed. So much so that in one of Coope's most famous discoveries, an assemblage of beetles from the 40 000-year-old carcass of a woolly rhinoceros from the Ukraine, 'the complete beetles were preserved down to the tarsal and antennal joints; . . . the wings could be unfolded and mounted; and parasitic mites, both larvae and adults, were found underneath the wings'. Even the insects' penises could be dissected and studied.

Over the years, Coope has come to suspect that glaciation hardly ever bisected beetle populations in Eurasia. Instead, the insects' ranges simply shifted southwards. For example, two closely related water beetles called *Helophorus aspericollis* and *H. brevicollis* nowadays occur in eastern Siberia and Europe, respectively. This gives the impression that they evolved during the last Ice Age, when vast glaciers separated these two areas. But Coope found the Siberian species in British fossil deposits dating back to the coldest part of the last Ice Age, whereas *brevicollis* took its place during an earlier warmer period. So whatever caused these species to split, it was not the glaciers. They were already there before that time, and wandered back and forth as climates got warmer or colder.

Birds do not make very good fossils, so when John Klicka and Robert Zink, two biologists of the University of Minnesota in St Paul, wanted to assess the relevance of Ice Ages to bird speciation, they took another approach. Many birds in the eastern part of North America have a west coast counterpart in the form of a related species or subspecies. For example, the well-known blue jay of eastern US city parks has a blue, white, and black mask, while on the west coast it is replaced by steller's jay, which has an all-black hood. Klicka and Zink identified 35 pairs of songbird species that were similarly divided into an eastern and a western form. Traditionally, ornithologists had always attributed the existence of these birds to the massive glaciers that had lain down the centre of North America during the latest Ice Age, between 250 000 and 10 000 years ago.

In the absence of fossils, the two biologists resorted to the birds' DNA to check if there was any truth in this assumption. As we saw before, geneticists can tell when two species split. They use the so-called 'molecular clock' of mitochondrial DNA. The rate with which this type of DNA mutates is roughly known—some 2% per million years. Nancy Knowlton used it to date the origin of her shrimps, and Klicka and Zink employed it as well. But unlike the Isthmus of Panama, the glaciers' age did not match the birds' DNA differences particularly well. Klicka and Zink found only a handful of bird species with DNA differences small enough for the populations to have been isolated by the glaciers. Indeed, most of the species had differences of around 6%, which meant the populations had split some 3 million years ago, before the Pleistocene ice ages started. As the two authors flatly state in their 1997 *Science* article, 'The entrenched paradigm that many North American songbird species originated as a consequence of these glaciations is flawed'.

Then what about islands? The word itself gave rise to the term 'isolation', so if there is any place where species have evolved in allopatry, it must be there. In a sense, this is correct. Many islands, especially those that are in the middle of an ocean, have their own species, which undoubtedly evolved there and nowhere else. Mauritius in the Indian Ocean boasts a motley crew of species that evolved from incidental vagrants that made it there. Hence, these endemics occur nowhere else in the world. The collection includes the Mauritius kestrel, the Mauritius banana fly, the Mauritius *Calvaria* tree, and, most famously, the now-extinct dodo.

But the case is less clear-cut for the more awe-inspiring productions of island speciation. The Hawaiian archipelago is home to almost 10 000 endemic species of plants and animals. Each one of them evolved there on the spot. The whole assemblage fanned out from just a small number of ancestors (a few hundred species at most) that once made their way to the islands. An immigrant finch gave rise, over millions of years, to some 50 species of honeycreepers. A Californian wildflower diversified into the 28 monumental silversword plants that dot the archipelago's volcanic slopes.

Although the link between speciation and the isolated bits of land in an archipelago seems obvious, it is not at all certain that it was the isolation that was responsible. When the Hawaiian Islands (or the Galápagos, or any other oceanic island group) sprang from the bottom of the sea, they were a clean slate, ready to be colonized by whatever the winds and the waves brought in. Once the ecosystem began to develop, niches began to appear that none of the present species could fill. Plants would grow with no specialized caterpillars to eat them. Flies would multiply without frogs to catch them. And fungi would grow unchecked by mushroom beetles. So just like the cichlid fish that discovered an uninhabited crater lake in Cameroon, or the fruit flies that found apples in Hudson River Valley waiting to be exploited, the Hawaiian immigrants rapidly diversified to fill all these vacant niches. They speciated on the islands not because there was geographical isolation, but because there was ecological opportunity. And as any entrepreneur knows, opportunity comes unexpectedly. When it's there, you have to grab it.

Sedimental journey

Evolution's path is littered with the traces of entrepreneurial species grabbing their ecological chances. These traces come in all sorts and sizes. They can be isolated events, like the host-shift of *Rhagoletis*; they can come in small clusters, like the nine cichlids in Lake Bermin in Cameroon; or in large clusters, as the 2000 different cichlids in East African lakes attest. But this is still nothing compared with what happened on that fateful day in 65 730 000 BC.

With a speed of 50 kilometres per second, a gigantic asteroid plunged into what is now the Yucatán Peninsula of Mexico, creating a 180-kilometre-wide crater and setting off earthquakes of magnitude 10 on the Richter scale in all directions. The heat generated by the impact was so severe, that within 10 minutes all forests in North America were ablaze, while debris from the impact started to rain down as thousands of mini-impacts. The same day, 50-metre-high sea waves called tsunamis began to slosh back and forth across the Atlantic, inundating the coasts of the Americas, Europe, and Africa. Dust thrown up into the atmosphere shrouded the planet in utter darkness and global temperatures plummeted. Photosynthesis came to a grinding halt. It took almost a year for the clouds to settle and 10 years for the temperature to creep back up. Meanwhile, the first extinctions in what is now known as the Cretaceous/Tertiary (or, in shorthand, K/T) mass extinction began to take place.

In the aftermath, many more species disappeared. The dinosaurs, which hung on for another 100 000 years and then vanished, were among the most famous victims. But shelled squid known as ammonites—now prominent display pieces in any fossil showcase—disappeared as well. Those that were not completely annihilated—the corals and sea lilies, for example—were brought back to just a shadow of their former selves: only small numbers of species remained. With as many as three-quarters of all species wiped out, Earth was as empty a place as it had been since times primordial.

As should now be apparent, speciation likes nothing better than a good empty niche. Given half a chance, it will rapidly fill it. And that is exactly what happened. Almost immediately after the great dying, new species started to appear, splitting off from the few species that had survived the cataclysm. Things happened slowly at first, but, as ecosystems began to reassemble, new niches opened up and a great profusion of speciation a few million years later brought species richness back to pre-catastrophe levels.

Speciation spasms as overwhelming as this one have been very rare in Earth's history. Only five cases of mass extinction followed by a biodiversity rebound are known. But smaller versions are much more common. As Michael Rosenzweig points out, speciation follows a so-called fractal pattern. Like the proverbial coastline that looks jagged

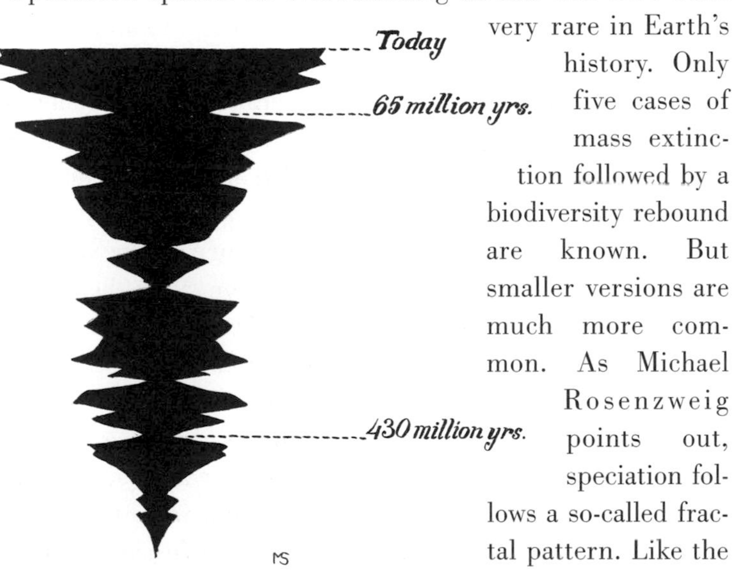

Bumpy ride: the number of species on Earth from 500 million years ago until today (extinctions and rebounds).

whichever map scale you look at, speciation is prompted by vacating niches at any scale between apple orchard and asteroid impact. In a few rare cases, what happened during such a smaller speciation spasm can be literally read from the rocks. In the Canadian province of Quebec, two palaeontologists called Peter and Sara Bretsky in the 1960s discovered a unique stack of ocean sediments dating back to the late Ordovician period, some 440 million years ago. The rocks record a time of great upheaval in the ecology of the ocean floor.

For the previous 100 million years, this particular environment hardly supported any animals: one species of trilobite, one or two species of graptolite (both extinct invertebrates), that was all. But in the late Ordovician, all that changed. Scientists are not quite sure what exactly happened. For some reason, the ocean floor suddenly became a very attractive place to live, possibly because oxygen concentrations there finally reached comfortable levels. The Quebec site, known as the Nicolet River Valley, records this opening up of niches with photographic precision. The 800-metre-tall flanks of the valley are like a gigantic hamburger made up of 218 layers, each separated by a few thousand years. Rosenzweig, who studied the deposits, says: 'It is like a sequence of 218 snapshots of the invertebrate fauna during some 5 million years.'

The most dramatic changeover can be seen in the very first snapshots. The initial four or five are typical for the meagre period before the niches opened up: none, one or two species present. But then the environment rapidly filled up. Successive snapshots show 5 species, then 10 species, 15, even 25. By the fortieth snapshot, less than a million years since the first, the fauna had reached its saturation point of some 40 species, which then kept stable for the remaining 4 million years (only to be wiped out by the mass extinction at the end of the Ordovician, which was every bit as big as the K/T event).

Where did all those new species come from? Most were snails, worms, and shelly things called brachiopods that were obviously unrelated to the few mud-inhabitants from earlier snapshots. And palaeontologists had never seen any of them before in older fossil beds. 'What appears to be going on', Rosenzweig says, 'is that occasionally an animal will be brought into the muddy ocean floor. And for the previous 100 million years it would die. But now all of a sudden it lives. It invades, and those invaders then speciate.' Speciation then must have been very quick. 'I can't even give you an estimate of how quickly it's gone', says Rosenzweig. 'I can't tell you whether it's all happened in 50 000 years or a 150 000 years, but it's something spectacularly fast like that.'

Fast tracks

'Spectacularly fast' is an expression not unfamiliar to palaeontologists. Time and time again, these scientists have seen new species appearing in the fossil record in a geological blink. In 1972, palaeontologists Niles Eldredge and Stephen Jay Gould even coined a special term for it. 'Punctuated equilibria' symbolized the palaeontological conundrum of long periods that species hardly changed at all, punctuated by sudden, almost instantaneous speciation.

Stephen Stanley, a palaeontologist at Johns Hopkins University in Baltimore, has been one of the chief flag-bearers of punctuated equilibria. During the 1970s he amassed evidence that species most of the time do not show any evolution. They can remain the same for millions and millions of years. Fossil layers from Wyoming, for example, show that the mammals that roamed there 40 million years ago, did so for periods of up to 5 million years without showing any evolutionary change. Almost all the species of mosses that were alive 20 million years ago are still alive today. And the tadpole shrimp *Triops cancriformis* beats them all. It survived untouched by evolution in Eurasian pools from 180 million years ago till this very day.

On the other hand, if there were changes in the fossils they were sudden and abrupt, as an article in the journal *Nature* of 8

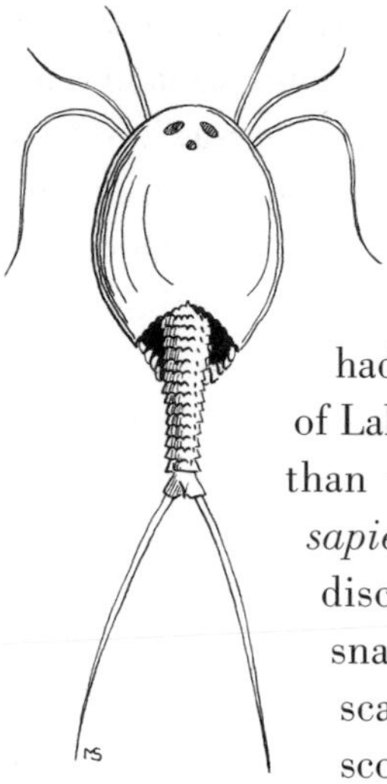

October 1981 showed. The paper, by the late Harvard University palaeontologist Peter Williamson, presented one of the most complete patterns of punctuated equilibria ever. Williamson had capitalized on the human fossil digs that had been going on in the ancient shorelines of Lake Turkana in northern Kenya. But rather than the media-sensitive forebears of *Homo sapiens*, Williamson concentrated on the more discreet inhabitants of the fossil layers: the snails and oysters. Freshwater shells were scattered in the layers of sediments by the scores. And like the Nicolet River site, Lake Turkana offered an uninterrupted view on 5 million years of evolution, with the added benefit that, here, the rocks had recorded the evolution of animals that were still thriving in the present-day lake.

The tadpole shrimp, *Triops cancriformis*, a species that has survived essentially unchanged for some 180 million years.

Williamson sampled and measured thousands of shells from 190 layers of sediment. When he started analysing his material, he discovered that species would remain unchanged for dozens of layers, spanning 3 million years or more and then suddenly, within only a few thousand years, split in two. A roundish snail called *Cleopatra ferruginea*, for example, suddenly gave rise to a species with a drawn-out tip, which persisted for a short while and then went extinct. At the same time, some 2.5 million years ago, sister species were spawned by another nine mollusc species at the erstwhile lake shore.

Williamson, like punctuated equilibria enthusiasts before him, attributed the punctuated equilibrium to Ernst Mayr's bottleneck model. The time that these speciations took place, he said, coincided with a period that geologists knew Lake Turkana had become much smaller and then much larger again. This gave the opportunity for the lake to be reconnected with smaller ones at its fringes. Any snails that had been evolving in isolation in those satellite lakes could now all of a sudden make their way into Lake Turkana and leave their shells in the fossil record. Williamson's model was not much different from the scenario that Humphry Greenwood had postulated for the Lake Victoria cichlids (see Chapter 2).

At the time the *Nature* paper came out, punctuated equilibria was the focus of a heated debate, which we will not get into here. But suffice it to say that the article made the 'punk eek war' reach an all time high. The following months, the pages of *Nature* regularly featured letters to the editor, to which Williamson then replied, only to be followed by more letters. The letters section of the 15 April 1982 issue of *Nature* supplied a generous helping of 'punk eek' correspondence. It featured a letter by Ernst Mayr himself, who had spotted other punctuations in Williamson's data and wanted those explained. It also featured letters by five other scientists from respectable universities who had some or other axe to grind. And then there was a tiny letter, less than half a column in length, by a certain D. W. Lindsay, writing from his home address in Maldon, Essex.

Why oh why, Lindsay innocently asked, was this entire debate devoid of any mention of natural selection? Surely the rise in lake level did more that just remove geographical isolation. It might have been the result of drastic climatic changes, Lindsay pondered, creating new niches and hence new species. 'We ought to consider a punctuated environment', he wrote. His letter was not really taken seriously. Williamson, in one of his replies, scoffed at Lindsay's suggestion, saying that it would be hard to believe that during millions of years the environment would have no effect on a species and then suddenly, in a few thousand years, it would. Anybody, Williamson claimed, could see that evolution was not just about 'simple fluctuations in selection pressures'. Instead, as was common (but, as scientists today think, largely flawed) knowledge, founder effects and genetic drift were what it was all about.

And yet, whoever he was, the timid correspondent in Maldon had hit the nail on the head. As it happened, the environment *had* been punctuated. The date, around 2.5–2.4 million years ago, heralded one of the cold and dry climatic cycles that had started on a global scale about half a million years previously. Before that time, the East African climate had been stable, warm, and wet. But from about 2.8 million years ago onwards, global weather was in perpetual indecision. First it cooled down, bringing the first Ice Age in the Northern Hemisphere. Water became locked up in the expanded polar caps, bringing dry spells to tropical regions. Rainforests and woods withdrew, and bushy grasslands came in their place. Then it would warm up again. And this cycle repeated itself many times over and over again, up to the present day.

The new Turkana snails had evolved at the onset of one of these climatic cycles. And they had not been the only African animals to have responded to the climate change by spawning new species. Elisabeth Vrba, a palaeontologist who was working at South Africa's Transvaal Museum in the early 1980s, was one of the first to recognize this. She studied African mammal fossils and realized that the time around 2.7–2.5 million years ago brought about large numbers of new species of antelope and horses, many of which evolved teeth adapted to eating the tough savannah grasses. Rodents, too, responded with a speciation pulse around that time.

Vrba, who is now at Yale University in New Haven, and extrapolating from antelopes to apes, thinks the changes to the African environment between 3 and 2 million years ago may also have been responsible for the evolution of several branches of our own family tree. During that period, first *Paranthropus aethiopicus* and *P. boisei* of eastern Africa evolved, and, later, *Paranthropus robustus* of southern Africa. These extinct 'bipedal apes' ('robust' upright walkers but side-shoots of the human family tree) had heavy-duty molars for grinding up a coarse vegetarian diet. The tops of their skulls bore large crests for anchoring massive jaw muscles. Another branch of the tree, with a more 'gracile' appearance and smaller chewing teeth, produced the meat-eating *Australopithecus garhi* and *A. africanus* and perhaps also *Homo*, our own genus. Ian Tattersall, a palaeontologist at the American Museum of Natural History in New York, has written that this period was one 'of intense speciation and evolutionary experimentation . . . It is from this ferment that the human ancestor ultimately emerged'.

What happened in Africa was echoed in other continents as well. Remember Klicka and Zink's twin species of American songbirds earlier in this chapter? Their DNA, too, showed most had split at roughly 2.5 million years ago. As Michael Rosenzweig has remarked: 'Curious indeed. The geographical barrier of the glacier itself does not create a ripple in speciation rate, but the climate change that preceded it does.' It may even have done more than that. The events that punctuated the world's environment between 3 and 2 million years ago, sending ripples of speciation across the globe, created the most unexpected niche of all in Africa. A niche for an animal that the world had not seen before. A meat-eater, who did funny things with pebbles. Its artefacts, the first human tools, turn up in geological layers dated around 2.5 million years old. And so do fossils of their supposed maker. The world would never be the same again.

Coda

Daybreak. Like a Chinese lantern, the sun rose perpendicularly from the still waters of the Pacific Ocean and cast its rays on to the eastern flanks of the volcanic island. The early morning light laid an ochre gloss over the deep green rainforest. A large bird rose from the branch where it had spent the night and lifted itself into the air with lazy, swooshing strokes. Small bees began to buzz around the flowers of the mangroves, while, on the beach, fiddler crabs made nervous excursions from perfectly round holes in the sand. A small group of them gathered around the carcass of a barracuda that had washed ashore during the night.

It was a morning like any other. The coral reefs thrived, the mountain streams gurgled, and the sprawling rainforest had once again reached all corners of the island, shooting the massive trunks of its 60-metre-high green-crowned emergents high up into the sky. It had been many millennia since the sound of the last chain-saw had died away. Only the rectangular rocks near the bay were a reminder of the large city that once had been here. Humans had come and gone, and the forest had repossessed the island, the string of islands in the distance, and even the one that had emerged from the sea not too long ago.

The sun rose higher and the light changed from yellow to bright white. The clouds that had clung to the steep slopes vanished, the bird choir fell silent, and the heat of mid-day sank like a humid blanket over the island. Only in the cooler, darker corners of the forest, some activity remained. Cicadas sang hysterically, a few large butterflies flapped about, and a bulky, black cockroach rummaged among the leaf-litter, looking for fallen fruit. Innately wary of the red junglebird that used to eat its ancestors, it kept a watchful eye at the specs of blue sky that peeked through the foliage overhead. It need not have done so. The red junglebird disappeared shortly after humans arrived, and it never returned.

Sadly, the cockroach had not yet evolved the awareness that, these days, danger came on foot. As it scurried around a fallen log, it was ruthlessly knocked down by the claw of a furry mammal. The cockroach felt itself being lifted from its feet, while sharp teeth started to nibble its abdomen, and the last thing its eyes registered was its predator's nose, unusually large, and spread out on the floor as a support.

Acknowledgements

This book has its roots in the time that I was a post-doc in the Department of Entomology at Wageningen University in the Netherlands. During that time, I led a sort of double life. During the day, I would be doing my research on parthenogenetic wasps, while in the evening I would write science stories for newspapers and magazines. Of course, some of my journalistic enterprises would spill over into my day job, which had me darting between my vials with DNA and the telephone where an impatient newspaper editor would be waiting.

It was on one of those hectic days that my room mate Greg Hurst, an Englishman with a deep affection for Dutch cows and other wonders of artificial and natural selection, turned his chair, rested his feet on my desk, and said: ‘Menno, why don’t you write a book about speciation?’ What he meant, of course, was: Why don’t you turn your writing into something that is really worthwhile.

One of my many weaknesses is that I tend to take advice from people too seriously, so when later that year I happened to meet Oxford University Press’s Lulu Stader at a congress in Budapest, I did my best to impress on her that a book on the evolution of new species was just what the world needed. Somehow, over the course of almost a year, I managed to convince her. She and Cathy Kennedy have sometimes forcefully, sometimes gently, but always dedicatedly coached me through this my first book project. Without them, I probably would have given up long ago. I also wish to thank Beth Knight and Sally Bunney for guiding me through the editing process.

Speciation is a vast subject and I have been able to get familiar with only a small subset of the thousands of publications. Worse still, I have had to make an even tinier selection for inclusion in this book. The choices I have made about which examples and theories to write about and which not, may not be to everybody's liking. I have excluded a lot of good and interesting science because of lack of space, or simply because a single example was often enough. Moreover, I probably have let my personal preferences for certain land animals prevail, which is why this book is somewhat poor in plants and marine creatures. I hereby apologize to them and to their aficionados.

Martin Enserink and Carol and Richard Stouthamer helped me in the early stages by reading sample chapters. Greg Hurst and his colleagues at University College London allowed me to hang around for a couple of weeks in January 1998 to soak up some literature in peace and quiet. My wife Sook-Peng Phoon and my friends Jeroen Roelfsema and Frank van Rooij read through large parts of the book as it neared completion and made me, as they always have, painfully but gratefully aware of my many errors and lapses of judgement. Other friends and colleagues who proof-read parts of the text are Arne Mooers, Anna Sand, Greg Hurst, Isabel Silva, Barbera Veldhuisen, Frans Zwanenburg, and two anonymous readers. I am very grateful to all of them for their time, effort, and their many thoughtful comments.

Many scientists were unbelievably patient and helped out by inviting me into their labs and homes, answering endless questions, providing unpublished manuscripts, reading through bits of draft, hinting at literature, having me interview them, or by giving the phone numbers and e-mail addresses of other scientists for me to pester. Among those I am most indebted to: Peter van Baarlen, Guy Bush, Peter van Dijk, John Endler, Claudia Englbrecht, Jeffrey Feder, Douglas Futuyma, Douglas Gill, Edi Gittenberger, Nancy Knowlton, Mark Macnair, Ernst Mayr, Steph Menken, Taka Miyatake, Anders Pape Møller, Arne Mooers, Bill Rice, Michael Rosenzweig, Uli Schliewen, Dolph Schluter, Ole Seehausen, Tom Smith, Pam Soltis, Diethard Tautz, and Tom Wood.

Other people and institutions who helped me in one way or another are: Jacques van Alphen, Takahiro Asami, Hans Breeuwer, Satoshi Chiba, Bill Eberhard, Frietson Galis, Sylvia Garagna, Carl Gerhardt, Theodoor Heijerman, Jody Hey, Katja Hora, Darren Irwin, Tom de Jong, Russ Lande, Jim Mallet, Hans Metz, Axel Meyer, Stewart Peck, Michel Perreau, Naomi Pierce, Trevor Price, Léon Raijmann, Peter Roessingh, Jeremy Searle, Chris Schneider, Alan Templeton, Bob Ursem, Wim Veldkamp, Bill Wcislo, James Weinberg, Jack Werren, Chung-I Wu, the library staff of Wageningen University, and my friends and colleagues at the Laboratory of Genetics there.

But, most of all, I thank Sook-Peng, Fenna, and Jan, and occasional guests to our house for putting up with my moods when I had writer's block, my brooding silence at the dinner table when I was thinking of the next line, my mourning when I had had to cut out an entire chapter, the closed door to my room, and my inexcusably frequent darting off to the library or to my office on bright and sunny summer days.

Menno Schilthuizen

Ede, the Netherlands
10 October 1999

Notes

PROLOGUE *The making of species* *(pp. 1–8)*

I acknowledge Quammen (1996) for making me aware of the strange world of tenrecs. A monograph on these mammals is Eisenberg and Gould (1970). The term 'divergence of character' is from Darwin (1859), while the term 'speciation', although first coined in 1906, came into general usage only as late as 1940 (Cook 1906; Cole 1940; see Berlocher 1998). The citation of Wilson is from Wilson (1992). The phrase 'Mystery of Mysteries' was actually coined by Darwin's contemporary, the astronomer John Herschel (1836; see Cannon 1961).

CHAPTER 1 *Sorting out life* *(pp. 1–31)*

If only it were true: The anecdote about Mayr in New Guinea is from Milner (1993).

Keep it in the family: For natural selection in grove snails, see Cain and Sheppard (1954). Mallet's quote is from Mallet (1997).

Into the mongrel zone: The percentages of hybridization in different groups come from Mallet (1995). The account of the fire-bellied toads is based on an interview with Jacek Szymura on June 21 1999 in Columbia, Missouri, and on Szymura (1993) and Szymura and Barton (1986, 1991). Linnaeus' description of the toads' sounds is from Blunt (1971). The bird hybrid zones in the American Great Plains are described by Moore and Price (1993). Cretan land-snail hybrid zones are described in Schilthuizen and Lombaerts (1994) and Schilthuizen (1995). For the hybrid zone of the hooded and carrion crows, see Meise (1928) and Palestrini and Rolando (1996).

BSc without honours: The account of snails on the Bonin Islands is based on Chiba (1993) and on conversations with Satoshi Chiba in August 1998, in Washington DC. The snails from Malaysian limestone hills are described in Tweedie (1961).

Back to Darwin: This section has been strongly influenced by Mallet (1995). Darwin's quotes are from Darwin (1845, 1859) and as cited in Milner (1993).

The age of genetics: Darwin's quote is from Darwin (1868). His ideas on the inheritance of tameness and other acquired traits were, however, not consistent, because in *On the Origin of Species* (Darwin 1859), he gives evidence against it. An animal with only one pair of chromosomes actually exists; an ant with such a minimalistic genome was reported by Crosland and Crozier (1986). The *Culex pipiens* work is described in Raymond *et al.* (1991).

Leaky genes: The tale of the jumping type C-virogene is from Benveniste (1985). The malaria-mosquito example is based on Garcia *et al.* (1996), Knols (1996), and an interview I had with Roger Butlin in Columbia, Missouri, on 20 June 1999. Mayr's quotes on co-adapted gene complexes are from Mayr (1963) while the quotes in the last paragraph are from a telephone interview I had with him on 16 June 1999.

When the gene pool freezes over: This chapter is based on an interview I had with Peter van Dijk and Peter van Baarlen on 19 May 1999 in Wageningen, the Netherlands, and also on Sterk *et al.* (1987).

CHAPTER 2 *An isolated case?* *(pp. 32–52)*

Darwin's quotes on the Galápagos Islands are all from Darwin (1845); his quote on Wagner is from Darwin (1872); and his note on Wagner's letter is mentioned in Sulloway (1979). Darwin was not against the notion of speciation by geographical isolation *per se*, but he was against the German's claim that species isolation was a prerequisite for evolution itself; according to Wagner, species would not change *unless* they became isolated.

The dew from heaven: I have taken the history of the discovery of birds of paradise, including the roles of Rothschild and Mayr, from Wallace (1869), Iredale (1950), Milner (1993), and Purcell and Gould (1993).

The giant wedge: The geological and natural history of the Isthmus of Panama is described in detail in Coates (1997). The account of Nancy Knowlton's work on snapping shrimps is based on Knowlton *et al.* (1993), Knowlton and Weigt (1998), and on a telephone interview I had with her on 24 September 1999.

Chasms cause schisms: The description of the underground laboratory at Moulis was based on an interview I had with Wim Veldkamp on 28 May 1999. A good text on cave life, on which the section on cave ecology was based, is Culver (1982). Darwin's quote is from Darwin (1859). The studies by the Juberthies are based on Juberthie-Jupeau (1988) and Juberthie (1988).

A ring of gulls: Martin Reid's website is at: http://www.martinreid.com/gullinx.htm. The information on the gull-ring species was taken from Mayr (1940, 1963) and Goethe (1955).

Fish on the edge: The general information on cichlids from the Great Lakes in Africa was compiled from numerous sources, including interviews and correspondence with Ole Seehausen (Leiden, 21 August 1998, February and March 1999), Axel Meyer's lecture in Leiden (15 November 1996), and, apart from the sources mentioned in the text, on Wilson (1992) and Goldschmidt (1996). It should be mentioned here that the way the Lake Nabugabo cichlids speciated is presumably not typical for Victoria cichlids. And the estimate for the lake's age (500 000 years) is probably a vast over-estimate (see Chapter 9).

CHAPTER 3 *Tight spots* *(pp. 53–72)*

Mayr's quote on the paradise kingfishers is from Mayr (1940), which is, together with Mayr (1954, 1963) the source for my account of his peripheral isolates model. A more recent treatment of paradise kingfishers can be found in Fry and Fry (1992). The founder effects in human populations have been taken from Milner (1993).

The great amalgamation: The major 1930s publications in which population genetics was founded are Fisher (1930), Wright (1931), and Haldane (1932). Dobzhansky's biography was taken from Lewontin (1997), Milner (1993), and Powell (1987). The major publications of the Modern Synthesis that are referred to in the text are Dobzhansky (1937), Huxley (1942), Mayr (1942), and Simpson (1944).

The creative force of a marginal existence: The outline of Mayr's peripheral isolates model is taken from Mayr (1954) and Mayr (1963), and his quotes are also taken from these two publications with the exception of 'perhaps . . . ever proposed', which I lifted from Quammen (1996). The punctuated equilibria theory was originally outlined in Eldredge and Gould (1972).

No, we have plenty bananas: My description of the diversity and geology of Hawaiian banana flies is largely based on Hollocher (1998) and Grant (1998).

Bottles and their necks: The biology of *Drosophila mercatorum* is taken from Templeton (1989) and from a lecture by him in Budapest, August 1996. His founder-flush experiments are described in Templeton (1979, 1989, 1996). The speciation event in laboratory worms is described in Weinberg *et al.* (1992).

Cracks in the bottles: Barton's critiques against bottleneck speciation are, for example, Barton (1989, 1998). The publication in which genetic diversity of Galápagos finches is assessed is Vincek *et al.* (1997). The Hawaiian *Drosophila* genetic variation is described in, for example, DeSalle *et al.* (1986) and DeSalle and Hunt (1987). The lab-worm was dethroned in Rodriguez-Trelles *et al.* (1996).

The anti-revolutionaries: The bottleneck experiments that could not convince Rice and Hostert (1993) are Powell (1978), Dodd and Powell (1985), Ringo *et al.* (1985), and Meffert and Bryant (1991). Templeton's reply to Rice and Hostert is Templeton (1996). Another critique of Templeton is Charlesworth *et al.* (1982).

The experiment to end all experiments: This section is based on an interview with Arne Mooers in Ede on 4 August 1999, and on Rundle *et al.* (1998) and Mooers *et al.* (1999). Mayr's quotes are from a telephone interview I had with him on 16 June 1999.

CHAPTER 4 *Seductive theories* *(pp. 73–99)*

In putting together this and the next section, I have been greatly helped by the clear and concise outline of the field of sexual selection in Ridley (1994). All quotes by Darwin are from Darwin (1871). The experiments to prove female choice in widowbirds, barn swallows, and great snipe are described in Andersson (1982), Møller (1988), and Höglund *et al.* (1990), respectively.

Good taste or good sense?: The work on mite infection in swallows is described in Møller (1990), while that on louse infection in grouse is from Boyce (1990) and Spurrier *et al.* (1991). The zebra finch experiments have been reported in Burley (1981) and Burley and Symanski (1998).

Speciation by fashion: Some of the more important theoretical papers are Møller and Pomiankowski (1993), Iwasa and Pomiankowski (1995), and Pomiankowski and Iwasa (1998).

Wondrous willies: The description of Malaysian sunbirds is from Jeyarajasingam and Pearson (1999). Flash patterns of American fireflies are illustrated in Lloyd (1966). Blenny courtship rituals can be found in Wirtz (1978). Some publications on the Zimbabwean banana fly and its courtship song are: Hollocher *et al.* (1997*a,b*) and Ritchie (1997). Butlin's remark is based on an interview I had with him in Columbia, Missouri, on 20 June 1999.

Polygamy breeds species: The paper referred to in this section is Møller and Cuervo (1998). The quotes by Møller are from a telephone interview I had with him on 23 August 1999. The Møller and Cuervo study is actually derived from earlier, similar publications, such as Barraclough *et al.* (1995) and Mitra *et al.* (1996).

The dark side of sex: Mating plugs, the traumatic copula of bedbugs, and the chemical composition of semen are discussed extensively in Eberhard (1985, 1996). The paragraph on sperm competition in banana flies was based on Price *et al.* (1999) and Birkhead (1999). The effect of male ejaculate on female health is taken from Chapman *et al.* (1995). The Red Queen hypothesis derives from Van Valen (1973), while the dance metaphor is taken from Rice (1998).

The seed of speciation: For the work on tsetse flies, see Machado (1959). The work with one tethered partner in *Drosophila*'s sexual dance is Rice (1996). Singh's work is in Thomas and Singh (1992) and Civetta and Singh (1995). The rodent testes gene is reported in Maiti *et al.* (1996). The quote by Roger Butlin is from an interview I had with him in Columbia, Missouri, on 20 June 1999. Rice's quotes are from Rice (1998) and from a telephone interview I had with him on 25 November 1998. The sterility of hybrid males between *Drosophila simulans* and *D. mauritiana* and one of the genes responsible are discussed in, for example, Ting *et al.* (1998).

Refurbishing geographical speciation: Mayr's notion that drongo and kingfisher plumage and insect genitalia shape were accidental byproducts of peripheral isolation is written down in Mayr (1963). The quote from Møller is from a telephone interview I had with him on 23 August 1999.

CHAPTER 5 *Wham, bam, brand new species* *(pp. 100–12)*

The outline of de Vries's life and work was taken from Milner (1993), the University of Amsterdam website, and a conversation with Bob Ursem on 21 September 1999. All the de Vries quotes come from de Vries (1901); the translations are from Milner (1993) or were done by myself. *Natura non facit Saltum* was used in Darwin (1859). The 1932 quote is by Clarence Ayres, T. H. Huxley's biographer, as quoted in Milner (1993). Details on the genetic specialities of evening primroses can be found in Dobzhansky (1970). Mayr's criticism on saltationism is based on Mayr (1963), while his dialogue with Goldschmidt was taken from Mayr (1980).

Double your genome today!: A good review of polyploid speciation in plants is Soltis and Soltis (1999). Ownbey's original publication is Ownbey (1950), while I have taken most other information from Novak *et al.* (1990) and from a telephone interview with Pam Soltis on 8 September 1999.

Please keep right: This section is based on Gittenberger (1988), Asami *et al.* (1998), conversations with Takahiro Asami in Washington DC in August, 1998, and an interview with Edi Gittenberger in Leiden on 3 September 1999. The original paper on the genetics of snail coiling is Boycott *et al.* (1930). The discussion on the simulatory likelihood of reversed-coil speciation is summarized as: Gittenberger (1988), Johnson *et al.* (1990), Orr (1991), and van Batenburg and Gittenberger (1996). The quoted alliteration is by Ken Zentzis, as uttered on the e-mail discussion list CONCH-L, on 9 September 1998.

CHAPTER 6 *A chronic case of* Rhagoletis *(pp. 113–30)*

The information on weevils is based on an interview with Theodoor Heijerman, in Wageningen, on 3 April 1998. The number of monophagous British leaf miners is taken from Strong *et al.* (1984). Terry Erwin's calculations are taken from Erwin (1982, 1983) and from a lecture by Erwin in Leiden, in 1988. Kevin Gaston's re-evaluation is from Gaston (1991).

A chronic case of Rhagoletis*:* Walsh's original publications are Walsh (1864, 1867). The account of Guy Bush's early years with Mayr and his quotes are taken from Bush (1998) and from a telephone interview I had with him on 21 April 1998.

Through the stomach: The paper by John Maynard Smith is Maynard Smith (1966). The account on the history of Bush's Ph.D. thesis and the quotes are, like in the last section, taken from Bush (1998) and from my telephone interview on 21 April 1998.

The miracle in the apple tree: Bush's original paper is Bush (1969). His speciation scenario was outlined in Bush (1975). The quotes by Doug Futuyma are based on an interview I had with him in Leiden on 27 March 1998. His paper is Futuyma and Mayer (1980), while Jaenike's is Jaenike (1981). The quotes by Guy Bush are based on a telephone interview I had with him on 21 April 1998.

Counter-offensive: The three 1988 papers in *Nature* are Feder *et al.* (1988), McPheron *et al.* (1988), and Smith (1988). The commentary in the same issue is Barton *et al.* (1988). Feder's papers are Feder *et al.* (1994, 1997). His quotes are from a telephone interview I had with him on 7 April 1998. Ron Prokopy's paper is Prokopy *et al.* (1988).

Joining up the dots—once again: The quotes by Douglas Futuyma are taken from an interview I had with him in Leiden on 27 March 1998, those from Mayr from a telephone interview on 16 June 1999. The report about the 1996 'Endless Forms' conference can be found in Gibbons (1996). The quotes from Feder and Bush are from telephone interviews I had with them on 7 and 21 April 1998, respectively. The account on the 'joining the dots' procedure is also based on the interview with Feder.

CHAPTER 7 *A freak show?* *(pp. 131–51)*

Mayr's quote is from a telephone interview I had with him on 16 June 1999. The account of host shifts in *Cydia pomonella* is based on Cisneros (1974) and Phillips and Barnes (1975). Further references to the host-race formation in moths on introduced plants on Hawaii and Rapa can be found in Strong *et al.* (1984), while the spider mite example was taken from Gotoh *et al.* (1993).

Caught in the act: The section on small ermine moths is based on a lecture by Léon Raijmann in Leiden on 18 December 1998, on an interview with Steph Menken in Amsterdam on 11 June 1998, and on Kooi *et al.* (1991), Raijmann (1996) and Menken and Roessingh (1998). Evidence for sympatric speciation in other insects can be found in Waring *et al.* (1990) and Craig *et al.* (1993) for goldenrod flies, in Hare and Kennedy (1986) for leaf beetles, and in Tauber and Tauber (1989) for lacewings. Treehoppers will be further discussed in Chapter 10.

A matter of taste: This section is largely based on an interview I had with Steph Menken in Amsterdam on 11 June 1998. Menken's quotes are also from that interview. Guy Bush's quotes are from a telephone interview I had with him on 21 April 1998. Bush's speciation model is described in Bush (1975) and the Hopkins effect was originally presented in Hopkins (1917).

Synchronized sex: In an altered form, this section also appeared in the 16 January 1999 issue of *New Scientist*. It is based on a conversation I had with Taka Miyatake in London on 16 January 1998, on an e-mail correspondence with him in December 1998, and on Miyatake and Shimizu (1999). The mating time table for giant silkmoths is based on Rau and Rau (1929), as cited in Wilson (1992). Rice's experiments are described in Rice and Salt (1988, 1990). His quotes are from a telephone interview I had with him on 25 November 1998, and from Rice and Hostert (1993).

Fishy business: This section is based on a visit to the Zoology Institute of the University of Munich, and interviews with Diethard Tautz and Uli Schliewen on 6 July 1998, and subsequent e-mail correspondence in January and February of 1999. Important publications on the Cameroon cichlids are Stiassny *et al.* (1992) and Schliewen *et al.* (1994). I also benefited greatly from an unpublished manuscript by Schliewen *et al.* Doebeli's computer simulations are in Doebeli (1996) and Dieckmann and Doebeli (1999).

Ice-Age lakes: The section on bullheads in the Königssee is based on conversations with Diethard Tautz, Uli Schliewen, and Claudia Englbrecht in Munich on 6 July 1998, and subsequent e-mail correspondence. Dolph Schluter's quotes are taken from a telephone interview I had with him on 29 January 1999. The section on sticklebacks is based on a lecture by Michael Bell in Leiden in 1993, and on Schluter (1994, 1996, 1998), Schluter and Nagel (1995), and Hatfield and Schluter (1999).

CHAPTER 8 *Ecotone—speciation prone* *(pp. 152–65)*

Huxley was not actually the first to emphasize that speciation could happen at ecotones. Fisher (1930) said the same. The quotes by Smith in this and the next section are from a telephone interview I had with him on 13 September 1999. In addition to the pleiotropy mentioned in this section, the build-up of sexual differences could also be reinforced if intermediates between the two populations would be less fit. In fact, this is what Endler envisaged. It is quite likely that the intermediates would do less well, because they would be adapted to neither of the two habitats. In such a case, natural selection would favour those animals with the most extreme mating signals, which would be least likely to mate with a trespasser from 'the other side'.

Birds of a feather won't stick together: The bit about Cameroonian greenbuls is based on Smith *et al.* (1997), Enserink (1997), and a telephone interview with Tom Smith on 13 September 1999. The work on sunbirds and finches is described in Smith *et al.* (in press).

The Pleistocene and the plasticine: This piece is based on Schneider *et al.* (1999), a telephone interview with Tom Smith on 13 September 1999, and e-mail correspondence with Chris Schneider.

Heavy metal and hard rock: The literature used for this section is Macnair (1983, 1987, 1989*a*,*b*), Macnair and Cumbes (1989), and Macnair and Gardner (1999). The quotes are from a telephone interview I had with Mark Macnair on 14 September 1999.

Back in the field: The snail work is described in, for example, Johannesson *et al.* (1995) and Tatarenkov and Johannesson (1998). The data on ecotones at cave entrances is based on Gertsch and Peck (1992), Peck and Finston

(1993), and on a telephone interview I had with Stewart Peck on 15 June 1999. More information on the Mexican cave fish can be found in, for example, Wilkens (1993), while the quotes from John Endler are taken from a telephone interview I had with him on 30 September 1999.

CHAPTER 9 *Victoria's blue genes* *(pp. 166–76)*

The account on the history of Lake Malawi's water levels is based on Goldschmidt (1996) and Owen *et al.* (1990). The genetics of the lake's cichlids is from van Oppen *et al.* (1997, 1998) and Reinthal and Meyer (1997). The Meyer quote is from a lecture he gave in Leiden on 15 November 1996.

Small fry: This section is based on conversations I had with Ole Seehausen in Leiden on 21 August 1998, and other occasions, and on Seehausen (1996, 1999), Goldschmidt (1996), and Seehausen *et al.* (1998).

The species pump: Again, the quotes are from an interview with Ole Seehausen in Leiden on 21 August 1998. The role of coloration in mate choice has been discussed in, for example, Seehausen *et al.* (1997), Seehausen and van Alphen (1998), and Seehausen (1999). George Turner's speciation model has been published as Turner and Burrows (1995). Further treatments of *mbipi* speciation are in Galis and Metz (1998). The two *Nature* papers are Dieckmann and Doebeli (1999) and Kondrashov and Kondrashov (1999). The new dating of Lake Victoria is in Johnson *et al.* (1996).

CHAPTER 10 *Mystery? What mystery?* *(pp. 177–97)*

Futuyma's quote is from Futuyma (1983). The report in *Science* about the 'Endless forms' congress is Gibbons (1996).

Treehoppers hop trees: This section is based on a telephone interview I had with Tom Wood in January 1999, and subsequent e-mail correspondence. References to Wood's work are Wood and Guttman (1983), Wood and Keese (1990), and Wood (1993).

Living apart together: The allopatric aspects of parasite speciation is further explored in Futuyma and Mayer (1980) and Brooks and McLennan (1993). The quotes by Nancy Knowlton are from a telephone interview I had with her on 24 September 1999. The Greek banana fly experiments are described in Kilias *et al.* (1980). Jeffrey Feder's quote is from a telephone interview I had with him on 7 April 1998.

What's sex got to do with it? The effect of predators in guppy coloration is from Endler (1980). The greater impact of parasites on tropical birds is from Møller (1998), and the work on cricket frog calls is in Ryan *et al.* (1990). Ernst Mayr's quotes are from a telephone interview I had with him on 16 June 1999. Mooers' quotes are from an interview in Ede on 4 August 1999.

Overrated barriers: The quote by Ernst Mayr is from Gibbons (1996). Lynch's work is in Lynch (1989). The quotes by Michael Rosenzweig come from Rosenzweig (1997), while Guy Bush's estimations of the impact of allopatry are from a telephone conversation I had with him on 21 April 1998. The quote by Nancy Knowlton is from a telephone interview on 24 September 1999. Coope's work and quotes are from Coope (1979), while those on American songbirds are from Klicka and Zink (1997). I took some information and views on the Hawaiian fauna from Schluter (1998).

Sedimental journey: The sequence of events after the Cretaceous/Tertiary impact are taken from Sepkoski (1999) and Conway Morris (1999). Rosenzweig's quote about fractality of speciation spasms is from Rosenzweig (1997), his other quotes are from a telephone interview I had with him on 6 October 1999. The information on the Nicolet River site is based on Rosenzweig and Taylor (1980), Rosenzweig (1995), and the interview.

Fast tracks: The original punctuated equilibria paper is Eldredge and Gould (1972). Some of Stanley's more important works are Stanley (1979, 1982), which is also where the examples mentioned are taken from. Williamson's primary paper is Williamson (1981). The chunk of subsequent debate including letters from Mayr, Lindsay, and Williamson himself are in *Nature*, Vol. 296 (15 April 1982), pp. 608-12. Theories on the impact of climate on evolution, especially mammalian and human speciation, are covered in *Paleoclimate and Evolution* edited by Vrba *et al.* (1995); with Chapters 3 and 27 by Vrba herself. I also used Milner (1993). The quote by Rosenzweig is from Rosenzweig (1997). I took the information on human biodiversity between 3 and 2 million years ago from Klein (2000) and Tattersall (2000).

References

Andersson, M. (1982). Female choice selects for extreme tail length in a widowbird. *Nature*, **299**, 818–20.

Asami, T., Cowie, R. H., and Ohbayashi, K. (1998). Evolution of mirror images by sexually asymmetric mating behavior in hermaphroditic snails. *The American Naturalist*, **152**, 225–36.

Barraclough, T. G., Harvey, P. H., and Nee, S. (1995). Sexual selection and taxonomic diversity in passerine birds. *Proceedings of the Royal Society of London*, **B259**, 211–15.

Barton, N. H. (1989). Founder effect speciation. In *Speciation and its Consequences*, (ed. D. Otte and J. A. Endler), pp. 229–56. Sinauer, Sunderland, MA.

Barton, N. H. (1998). Natural selection and random genetic drift as causes of evolution on islands. In *Evolution on Islands*, (ed. P. R. Grant), pp. 102–23. Oxford University Press, Oxford.

Barton, N. H., Jones, J. S., and Mallet, J. (1988). No barriers to speciation. *Nature*, **336**, 13–14.

Batenburg, F. H. D. van and Gittenberger, E. (1996). Ease of fixation of a change in coiling: computer experiments on chirality in snails. *Heredity*, **76**, 278–86.

Bateson, W. (1922). Evolutionary faith and modern doubts. *Science* **55**, 55–61.

Benveniste, R. E. (1985). The contributions of retroviruses to the study of mammalian evolution. In *Molecular Evolutionary Genetics*, (ed. R. I. MacIntyre), pp. 359–417. Plenum, New York.

Berlocher, S. H. (1998). Origins—a brief history of research on speciation. In *Endless Forms: Species and Speciation*, (ed. D. J. Howard and S. H. Berlocher), pp. 3–15. Oxford University Press, Oxford.

Birkhead, T. (1999). Distinguished sperm in competition. *Nature*, **400**, 406–7.

Blunt, W. (1971). *A Life of Linnaeus*. Collins, London.

Boyce, M. S. (1990). The Red Queen visits sage grouse leks. *The American Naturalist*, **30**, 263–70.

Boycott, A. E., Diver, C., Garstang, S. L., and Turner, F. M. (1930). The inheritance of sinistrality in *Limnaea peregra* (Mollusca, Pulmonata). *Transactions of the Royal Society of London*, **B219**, 51–131.

Bretsky, P. W. and Bretsky, S. S. (1975). Succession and repetition in Late Ordovician fossil assemblages from the Nicolet River valley, Quebec. *Paleobiology*, **1**, 225–37.

Bretsky, P. W. and Bretsky, S. S. (1976). The maintenance of evolutionary equilibrium in Late Ordovician benthic marine invertebrate faunas. *Lethaia*, **9**, 223–33.

Brooks, D. R. and McLennan, D. A. (1993). *Parascript: Parasites and the Language of Evolution*. Smithsonian Institution Press, Washington.

Burley, N. T. (1981). Sex ratio manipulation and selection for attractiveness. *Science*, **211**, 721–2.

Burley, N. T. and Symanski, R. (1998). 'A taste for the beautiful': latent aesthetic mate preferences for white crests in two species of Australian grass finches. *The American Naturalist*, **152**, 792–802.

Bush, G. L. (1969). Sympatric host race formation and speciation in frugivorous flies of the genus *Rhagoletis* (Diptera, Tephritidae). *Evolution*, **23**, 237–51.

Bush, G. L. (1975). Modes of animal speciation. *Annual Review of Ecology and Systematics*, **6**, 339–64.

Bush, G. L. (1994). Sympatric speciation in animals—new wine in old bottles. *Trends in Ecology and Evolution*, **9**, 285–8.

Bush, G. L. (1998). The conceptual radicalization of an evolutionary biologist. In *Endless Forms: Species and Speciation*, (ed. D. J. Howard and S. H. Berlocher), pp. 425–38. Oxford University Press, Oxford.

Cannon, W. F. (1961). The impact of uniformitarianism. *Proceedings of the American Philosophical Society*, **105**, 301–14.

Cain, A. J. and Sheppard, P. M. (1954). Natural selection in *Cepaea*. *Genetics*, **39**, 89–116.

Carson, H. L. (1968). The population flush and its genetic consequences. In *Population Biology and Evolution*, (ed. R. C. Lewontin), pp. 123–37. Syracuse University Press, New York.

Chapman, T., Liddle, L. F., Kalb, J. M., Wolfner, M. F., and Partridge, L. (1995). Cost of mating in *Drosophila melanogaster* is mediated by male accessory gland products. *Nature*, **373**, 241–4.

Charlesworth, B., Lande, R., and Slatkin, M. (1982). A neo-Darwinian commentary on macroevolution. *Evolution*, **36**, 474–98.

Chiba, S. (1993). Modern and historical evidence for natural hybridization between sympatric species in *Mandarina* (Pulmonata: Camaenidae). *Evolution*, **47**, 1539–56.

Cisneros, F.H. (1974). Contribution to the biological and ecological characterization of apple and walnut host races of codling moth, *Laspeyresia pomonella* (L.): moth longevity and oviposition capacity. *Environmental Entomology*, **3**, 402–6.

Civetta, A. and Singh, R. S. (1995). High divergence of reproductive tract proteins and their association with post-zygotic reproductive isolation in *Drosophila melanogaster* and *Drosophila virilis* group species. *Journal of Molecular Evolution*, **41**, 1085–95.

Coates, A. G. (1997). *Central America: A Natural and Cultural History*. Yale University Press, New Haven.

Cole, L. J. (1940). The relation of genetics to geographical distribution and speciation; speciation I. Introduction. *The American Naturalist*, **74**, 193–7.

Conway Morris, S. (1999). The evolution of diversity in ancient ecosystems: a review. In *Evolution of Biological Diversity*, (ed. A. E. Magurran and R. M. May), pp. 283–321. Oxford University Press, Oxford.

Cook, O. F. (1906). Factors of species-formation. *Science*, **23**, 506–7.

Coope, G. R. (1979). Late Cenozoic fossil Coleoptera: evolution, biogeography, and ecology. *Annual Review of Ecology and Systematics*, **10**, 247–67.

Craig, T. P., Itami, J. K., Abrahamson, W. G., and Horner, J. D. (1993). Behavioral evidence for host-race formation in *Eurosta solidaginis*. *Evolution*, **47**, 1696–1710.

Crosland, M. J. and Crozier, R. H. (1986). *Myrmecia pilosula*, an ant with only one pair of chromosomes. *Science*, **231**, 1278.

Culver, D. (1982). *Cave Life: Evolution and Ecology*. Harvard University Press, Cambridge, MA.

Darwin, C. (1842). *The Structure and Distribution of Coral Reefs*. Smith, Elder, London.

Darwin, C. (1845). *Journal of Researches into the Geology and Natural History of the Various Countries Visited by HMS* Beagle, *under the Command of Captain Fitzroy, R.N. from 1832 to 1836*, (2nd revised edn). Henry Colburn, London (abridged text, incorporating Fitzroy's appendices and with new introduction and notes published by Penguin Books, London, 1989).

Darwin, C. (1859). *On the Origin of Species by Means of Natural*

Selection, or the Preservation of Favoured Races in the Struggle for Life. Murray, London.

Darwin, C. (1868). *The Variation of Animals and Plants under Domestication*. Murray, London.

Darwin, C. (1871). *The Descent of Man and Selection in Relation to Sex*. Murray, London.

Darwin, C. (1872). *On the Origin of Species by Means of Natural Selection, or the Preservation of Favoured Races in the Struggle for Life*, (6th edn). Murray, London.

DeSalle, R. and Hunt, J. A. (1987). Molecular evolution in Hawaiian Drosophilids. *Trends in Ecology and Evolution*, **2**, 212–16.

DeSalle, R., Giddings, L. V., and Templeton, A. R. (1986). Mitochondrial DNA variability in natural populations of Hawaiian *Drosophila*. I. Methods and levels of variability in *D. sylvestris* and *D. heteroneura*. *Heredity*, **56**, 75–86.

Dieckmann, U. and Doebeli, M. (1999). On the origin of species by sympatric speciation. *Nature*, **311**, 354–7.

Dobzhansky, Th. (1937). *Genetics and the Origin of Species*. Columbia University Press, New York.

Dobzhansky, Th. (1970). *Genetics and the Evolutionary Process*. Columbia University Press, New York.

Dodd, D. M. B. and Powell, J. R. (1985). Founder-flush speciation: an update of experimental results with *Drosophila*. *Evolution*, **39**, 1388–92.

Doebeli, M. (1996). A quantitative genetic competition model for sympatric speciation. *Journal of Evolutionary Biology*, **9**, 893–909.

Eberhard, W. G., (1985). *Sexual Selection and Animal Genitalia*. Harvard University Press, Cambridge, MA.

Eberhard, W. G. (1996). *Sexual Selection by Cryptic Female Choice*. Princeton University Press, Princeton.

Eisenberg, J. F. and Gould, E. (1970). *The Tenrecs: A Study in Mammalian Behaviour and Evolution*, Smithsonian Contributions to Zoology, No. 27. Smithsonian Institution Press, Washington, DC.

Eldredge, N. and S. J. Gould (1972). Punctuated equilibria: an alternative to phyletic gradualism. In *Models in Paleobiology*, (ed. T. J. M. Schopf), pp. 82–115. Freeman, Cooper, San Francisco.

Endler, J. A. (1977). *Geographic Variation, Speciation, and Clines*. Princeton University Press, Princeton.

Endler, J. A. (1980). Natural selection on color patterns in *Poecilia reticulata*. *Evolution*, **34**, 76–91.

Enserink, M. (1997). Life on the edge: rainforest margins may spawn species. *Science*, **276**, 1791–2.

Erwin, T. L. (1982). Tropical forests: their richness in Coleoptera and other arthropod species. *Coleopterists Bulletin*, **36**, 74–5.

Erwin, T. L. (1983). Beetles and other insects of tropical forest canopies at Manaus, Brazil, sampled by insecticidal fogging. In *Tropical Rain Forest: Ecology and Management*, (ed. S. L. Sutton, T. C. Whitmore, and A. C. Chadwick), pp. 59–75. Blackwell, London.

Feder, J. L., Chilcote, C. A., and Bush, G. L. (1988). Genetic differentiation between sympatric host races of the apple maggot fly *Rhagoletis pomonella*. *Nature*, **336**, 61–4.

Feder, J. L., Opp, S., Wlazlo, B., Reynolds, K., Go, W., and Spisak, S. (1994). Host fidelity is an effective pre-mating barrier between sympatric races of the apple maggot fly. *Proceedings of the National Academy of Sciences USA*, **91**, 7990–4.

Feder, J. L., Roethele, J. B., Wlazlo, B., and Berlocher, S. H. (1997). The selective maintenance of allozyme differences between sympatric host races of the apple maggot fly. *Proceedings of the National Academy of Sciences USA*, **94**, 11417–21.

Fisher, R. A. (1930). *The Genetical Theory of Natural Selection*. Clarendon Press, Oxford.

Fry, C. H. and Fry, K. (1992). *Kingfishers, Bee-eaters and Rollers: A Handbook*. Princeton University Press, Princeton.

Futuyma, D. J. (1983). Mechanisms of speciation. *Science*, **219**, 1059–60.

Futuyma, D. J. and Mayer, G. C. (1980). Non-allopatric speciation in animals. *Systematic Zoology*, **29**, 254–71.

Galis, F. and Metz, J. A. J. (1998). Why are there so many cichlid species? *Trends in Ecology and Evolution*, **13**, 1–2.

Garcia, B. A., Caccone, A., Mathiopoulos, K. D., and Powell, J. R. (1996). Inversion monophyly in African anopheline malaria vectors. *Genetics*, **143**, 1313–20.

Gaston, K. J. (1991). The magnitude of global insect species richness. *Conservation Biology*, **5**, 283–96.

Gertsch, W. J. and Peck, S. B. (1992). The pholcid spiders of the Galápagos Islands, Ecuador (Araneae: Pholcidae). *Canadian Journal of Zoology*, **70**, 1185–99.

Gibbons, A. (1996). On the many origins of species. *Science*, **273**, 1496–8.

Gittenberger, E. (1988). Sympatric speciation in snails; a largely neglected model. *Evolution*, **42**, 826–8.

Goethe, F. (1955). Vergleichende Beobachtungen zum Verhalten der Silbermöwe (*Larus argentatus*) und der Heringmöwe (*Larus fuscus*). *Congress of Ornithology*, **37**, 577–82.

Goldschmidt, R. (1940). *The Material Basis of Evolution*. Yale University Press, New Haven.

Goldschmidt, T. (1996). *Darwin's Dreampond*. MIT Press, Cambridge, MA.

Gotoh, T., Bruin, J., Sabelis, M. W., and Menken, S. B. J. (1993). Host race formation in *Tetranychus urticae*: genetic differentiation, host plant preference, and mate choice in a tomato and a cucumber strain. *Entomologia Experimentalis et Applicata*, **68**, 171–8.

Grant, P. R. (1998). Speciation. In *Evolution on Islands*, (ed. P. R. Grant), pp. 83–101. Oxford University Press, Oxford.

Grant, V. (1963). *The Origin of Adaptations*. Columbia University Press, New York.

Greenwood, P. H. (1965). The cichlid fishes of Lake Nabugabo, Uganda. *Bulletin of the British Museum (Natural History) (Zoology)*, **12**, 315–57.

Haldane, J. B. S. (1932). *The Causes of Evolution*. Harper, London.

Hare, J. D. and Kennedy, G. G. (1986). Genetic variation in plant–insect associations: survival of *Leptinotarsa decemlineata* populations on *Solanum carolinense*. *Evolution*, **40**, 1031–43.

Hatfield, T. and Schluter, D. (1999). Ecological speciation in sticklebacks: environment-dependent hybrid fitness. *Evolution*, **53**, 866–73.

Höglund, J., Eriksson, M., and Lindell, L. E. (1990). Females of the lek-breeding great snipe, *Gallinago media*, prefer males with white tails. *Animal Behaviour*, **40**, 23–32.

Hollocher, H. (1998). Island hopping in *Drosophila*: genetic patterns and speciation mechanisms. In *Evolution on Islands*, (ed. P. R. Grant), pp. 124–41. Oxford University Press, Oxford.

Hollocher, H., Ting, C.-T., Pollack, F., and Wu, C.-I. (1997*a*). Incipient speciation by sexual isolation in *Drosophila melanogaster*: variation in mating preferences and correlation between sexes. *Evolution*, **51**, 1175–81.

Hollocher, H., Ting, C.-T., Wu, M.-L., and Wu, C.-I. (1997*b*). Incipient speciation by sexual isolation in *Drosophila melanogaster*: extensive genetic divergence without reinforcement. *Genetics*, **147**, 1191–201.

Hopkins, A. D. (1917). Entomologists' discussions. *Journal of Economic Entomology*, **10**, 92–3.

Hoshizaki, S. and Shimada, T. (1995). PCR-based detection of *Wolbachia*, cytoplasmic incompatibility microorganisms in natural populations of *Laodelphax striatellus* (Homoptera: Delphacidae) in central Japan: has the distribution of *Wolbachia* spread recently? *Insect Molecular Biology*, **4**, 237–43.

Huxley, J. S. (1939). Clines: an auxiliary method in taxonomy. *Bijdragen tot de Dierkunde*, **27**, 491–520.

Huxley, J. S. (1942). *Evolution: The Modern Synthesis*. Allen & Unwin, London.

Hutchinson, G. E. (1959). Homage to Santa Rosalia or why are there so many kinds of animals? *The American Naturalist*, **93**, 145–59.

Iredale, T. (1950). *Birds of Paradise and Bower Birds*. Georgian House, Melbourne.

Iwasa, Y. and Pomiankowski, A. (1995). Continual change in mate preferences. *Nature*, **377**, 420–2

Jaenike, J. (1981). Criteria for ascertaining the existence of host races. *American Naturalist* **117**, 830–4.

Janzen, D. (1988). Ecological characterization of a Costa Rican dry forest caterpillar fauna. *Biotropica*, **20**, 120–35.

Jeyarajasingam, A. and Pearson, A.(1999). *A Field Guide to the Birds of West Malaysia and Singapore*. Oxford University Press, Oxford.

Johannesson, K., Rolán-Alvarez, E., and Ekendahl, A. (1995). Incipient reproductive isolation between two sympatric morphs of the intertidal snail *Littorina saxatilis*. *Evolution*, **49**, 1180–90.

Johnson, M. S., Clarke, B., and Murray, J. (1990). The coil polymorphism in *Partula suturalis* does not favor sympatric speciation. *Evolution*, **44**, 459–64.

Johnson, T. C., Scholtz, C. A., Talbot, M. R., Kelts, K., Ricketts, R. D., Ngobi, G. *et al.* (1996). Late Pleistocene desiccation of Lake Victoria and rapid evolution of cichlid fishes. *Science*, **273**, 1091–3.

Juberthie, C. (1988). Paleoenvironment and speciation in the cave beetle complex *Speonomus delarouzeei* (Coleoptera, Bathysciinae). *International Journal of Speleology*, **17**, 31–50.

Juberthie-Jupeau, L. (1988). Mating behaviour and barriers to hybridization in the cave beetle of the *Speonomus delarouzeei* complex (Coleoptera, Catopidae, Bathysciinae). *International Journal of Speleology*, **17**, 51–63.

Kilias, G., Alahiotis, S. N., and Pelecanos, N. (1980). A multifactorial genetic investigation of speciation theory using *Drosophila melanogaster*. *Evolution*, **34**, 730–7.

Klein, R. G. (2000). Archeology and the evolution of human behavior. *Evolutionary Anthropology*, **9**, 17–36

Klicka, J. and Zink, R. M. (1997). The importance of recent ice ages in speciation: a failed paradigm. *Science*, **277**, 1666–9.

Knols, B. G. J. (1996). Odour-mediated host-seeking behaviour of the Afro-tropical malaria vector *Anopheles gambiae* Giles. Unpublished Ph.D. thesis. Wageningen Agricultural University, Wageningen.

Knowlton, N. and Weigt, L. A. (1998). New dates and new rates for divergence across the Isthmus of Panama. *Proceedings of the Royal Society of London*, **B265**, 2257–63.

Knowlton, N., Weigt, L. A., Solórzano, L. A., Millis, D. K., and Bermingham, E. (1993). Divergence in proteins, mitochondrial DNA, and reproductive compatibility across the Isthmus of Panama. *Science*, **260**, 1629–32.

Kondrashov, A. S. and Kondrashov, F. A.(1999). Interactions among quantitative traits in the course of sympatric speciation. *Nature*, **311**, 351–53.

Kooi, R. E., van de Water, T. P. M., and Herrebout, W. M. (1991). Host–plant selection and larval food acceptance by *Yponomeuta padellus*. *Proceedings of the Koninklijke Nederlandse Akademie van Wetenschappen* **94**, 221–32.

Lessios, H. A. (1998). The first stage of speciation as seen in organisms separated by the Isthmus of Panama. In *Endless Forms: Species and Speciation*, (ed. D. J. Howard and S. Berlocher), pp. 186–201. Oxford University Press, Oxford.

Lewontin, R. C. (1974). *The Genetic Basis of Evolutionary Change*. Columbia University Press, New York.

Lewontin, R. C. (1997). Dobzhansky's *Genetics and the Origin of Species*: is it still relevant? *Genetics*, **147**, 351–5.

Lloyd, J. E. (1966). Studies on the flash communication system in *Photinus* fireflies. *Miscellaneous Publications of the Museum of Zoology of the University of Michigan*, **130**, 1–195.

Lynch, J. D. (1989). The gauge of speciation: on the frequencies of modes of speciation. In *Speciation and its Consequences*, (ed. D. Otte and J. A. Endler), pp. 527–53. Sinauer, Sunderland, MA.

Machado, A. de Barros (1959). Nouvelles contributions à l'étude systématique et biogéographique des Glossines (Diptera). *Publicações Culturais da Companhia de Diamantes de Angola, Lisbon*, **46**, 13–90.

Macnair, M. R. (1983). The genetic control of copper tolerance in the yellow monkey flower, *Mimulus guttatus*. *Heredity*, **50**, 283–93.

Macnair, M. R. (1987). Heavy metal tolerance in plants: a model evolutionary system. *Trends in Ecology and Evolution*, **2**, 354–59.

Macnair, M. R. (1989*a*). A new species of *Mimulus* endemic to copper mines in California (USA). *Botanical Journal of the Linnean Society*, **100**, 1–14.

Macnair, M. R. (1989*b*). The potential for rapid speciation in plants. *Genome*, **31**, 203–10.

Macnair, M. R. and Cumbes, Q. J. (1989). The genetic architecture of interspecific variation in *Mimulus*. *Genetics*, **122**, 211–22.

Macnair, M. R. and Gardner, M. (1999). The evolution of edaphic endemics. In *Endless Forms: Species and Speciation*, (ed. D. J. Howard and S. H. Berlocher), pp. 157–71. Oxford University Press, Oxford.

Maiti, S., Doskow, J., Sutton, K., Nhim, R. P., Lawlor, D. A., Levan, K., *et al.* (1996). The *Pem* homeobox gene: rapid evolution of the homeodomain, X-chromosomal localization, and expression in reproductive tissue. *Genomics*, **34**, 304–16.

Mallet, J. (1995). A species definition for the Modern Synthesis. *Trends in Ecology and Evolution*, **10**, 294–99.

Mallet, J. (1997). What are species? *Trends in Ecology and Evolution*, **12**, 453–4.

May, R. M. (1988). How many species are there on earth? *Science*, **241**, 1441–9.

May, R. M. (1990). How many species? *Philosophical Transactions of the Royal Society of London*, **B330**, 293–304.

Maynard Smith, J. (1966). Sympatric speciation. *The American Naturalist*, **104**, 487–90.

Mayr, E. (1940). Speciation phenomena in birds. *The American Naturalist*, **74**, 249–78.

Mayr, E. (1942). *Systematics and the Origin of Species*. Columbia University Press, New York.

Mayr, E. (1954). Change of genetic environment and evolution. In *Evolution as a Process*, (ed. J. Huxley, A. C. Hardy, and E.B. Ford), pp. 157–80. Allen & Unwin, London.

Mayr, E. (1963). *Animal Species and Evolution*. Harvard University Press, Cambridge, MA.

Mayr, E. (1980). Prologue: some thoughts on the history of the evolutionary synthesis. In *The Evolutionary Synthesis*, (ed. E. Mayr and W. Provine), pp. 1–48. Harvard University Press, Cambridge, MA.

Mayr, E. (1992). Controversies in retrospect. *Oxford Surveys in Evolutionary Biology*, **8**, 1–34.

McCune, A. R. and Lovejoy, N. R. (1998). The relative rate of sympatric and allopatric speciation in fishes. In *Endless Forms: Species and Speciation*, (ed. D. J. Howard and S. H. Berlocher), pp. 172–85. Oxford University Press, Oxford.

McPheron, B. A., Smith, D. C., and Berlocher, S. H. (1988). Genetic differences between *Rhagoletis pomonella* host races. *Nature*, **336**, 64–6.

Meffert, L. M. and Bryant, E. H. (1991). Mating propensity and courtship behavior in serially bottlenecked lines of the housefly. *Evolution*, **45**, 293–306.

Meise, W. (1928). Die Verbreitung der Aaskrähe (Formenkreis *Corvus corone* L.). *Journal für Ornithologie*, **76**, 1–203.

Meisenheimer, J. (1912). Die Weinbergschnecke *Helix pomatia* L. In *Monographien einheimischer Tiere*, (ed. H. E. Ziegler and R. Wolterek), pp. 1–140. Werner Klinkhardt, Leipzig.

Menken, S. B. J. and Roessingh, P. (1998). Evolution of insect-plant associations: sensory perception and receptor modifications, direct food specialization, and host shifts in phytophagous insects. In *Endless Forms:Species and Speciation*, (ed. D. J. Howard and S. H. Berlocher), pp. 145–56. Oxford University Press, Oxford.

Meyer, A., Kocher, T. D., Basasibwaki, P., and Wilson, A. C. (1990). Monophyletic origin of Lake Victoria cichlid fishes suggested by mitochondrial DNA sequences. *Nature*, **347**, 550–3.

Milner, R. (1993). *The Encyclopedia of Evolution: Humanity's Search for its Origins*. Henry Holt, New York.

Mitra, S., Landel, H., and Pruett-Jones, S. (1996). Species richness covaries with mating system in birds. *Auk*, **113**, 544–51.

Miyatake, T. and Shimizu, T. (1999). Genetic correlations between life-history and behavioral traits can cause reproductive isolation. *Evolution*, **53**, 201–8.

Møller, A. P. (1988). Female choice selects for male sexual trait ornaments in the monogamous swallow. *Nature*, **332**, 640–2.

Møller, A. P. (1990). Effects of a haematophagous mite on secondary sexual tail ornaments in the barn swallow (*Hirundo rustica*): a test of the Hamilton and Zuk hypothesis. *Evolution*, **44**, 771–84.

Møller, A. P. (1998). Evidence of larger impact of parasites on hosts in the tropics: investment in immune function within and outside the tropics. *Oikos*, **82**, 265–70.

Møller, A. P. and Cuervo, J. J. (1998). Speciation and feather ornamentation in birds. *Evolution*, **52**, 859–69.

Møller, A. P. and Pomiankowski, A. (1993). Punctuated equilibria or gradual evolution: fluctuating asymmetry and variation in the rate of evolution. *Journal of Theoretical Biology*, **161**, 359–67.

Mooers, A. Ø., Rundle, H. D., and Whitlock, M. C. (1999). The effects of selection and bottlenecks on male mating success in peripheral isolates. *The American Naturalist*, **153**, 437–44.

Moore, W. S. and Price, J. T. (1993). Nature of selection in the Northern Flicker hybrid zone and its implications for speciation theory. In *Hybrid Zones and the Evolutionary Process*, (ed. R. G. Harrison), pp. 196–225. Oxford University Press, Oxford.

Mousson, A. (1849). *Die Land- und Süsswasser-Mollusken von Java.* Friedrich Schultess, Zürich.

Novak, S. J., Soltis, D. E., and Soltis, P. S. (1990). Ownbey's Tragopogons: 40 years later. *American Journal of Botany*, **78**, 1586–600.

Oppen, M. J. H. van, Turner, G. F., Rico, C., Deutsch, J. C., Ibrahim, K. M., Robinson, R. L., and Hewitt, G. M. (1997). Unusually fine-scale genetic structuring found in rapidly speciating Malawi cichlid fishes. *Proceedings of the Royal Society of London* **B264**, 1803–12.

Oppen, M. J. H. van, Turner, G. F., Rico, C., Robinson, R. L., Deutsch, J. C., Genner, M. J., and Hewitt, G. M. (1998). Assortative mating among rock-dwelling cichlid fishes supports high estimates of species richness from Lake Malawi. *Molecular Ecology*, **7**, 991–1001.

Orr, H. A. (1991). Is single-gene speciation possible? *Evolution*, **45**, 764–9.

Owen, R. B., Crossley, R., Johnson, T. C., Tweddle, D., Kornfield, I., Davison, S., *et al.* (1990). Major low levels of Lake Malawi and their implications for speciation rates in cichlid fishes. *Proceedings of the Royal Society of London*, **B240**, 519–53.

Ownbey, M. (1950). Natural hybridization and amphiploidy in the genus *Tragopogon*. *American Journal of Botany*, **37**, 485–99.

Palestrini, C. and Rolando, A. (1996). Differential calls by carrion and hooded crows (*Corvus cornone corone* and *C. c. cornix*) in the Alpine hybrid zone. *Bird Study*, **43**, 364–70.

Peck, S. B. and Finston, T. L. (1993). Galápagos Islands troglobites: the questions of tropical troglobites, parapatric distributions with eyed-sister-species, and their origin by parapatric speciation. *Mémoires de Biospéologie*, **20**, 19–37.

Phillips, P. A. and Barnes, M. M. (1975). Host race formation among sympatric apple, walnut, and plum populations of the codling moth,

Laspeyresia pomonella. *Annals of the Entomological Society of America*, **68**, 1053–60.

Pomiankowski, A. and Iwasa, Y. (1998). Runaway ornament diversity caused by Fisherian sexual selection. *Proceedings of the National Academy of Sciences USA*, **95**, 5106–11.

Powell, J. R. (1978). The founder-flush speciation theory: an experimental approach. *Evolution*, **32**, 465–74.

Powell, J. R. (1987). 'In the air'—Theodosius Dobzhansky's *Genetics and the Origin of Species*. *Genetics*, **117**, 363–6.

Powell, J. R. (1997). *Progress and Prospects in Evolutionary Biology: The* Drosophila *Model*. Oxford University Press, Oxford.

Price, C. S. C., Dyer, K. A., and Coyne, J. A. (1999). Sperm competition between *Drosophila* males involves both displacement and incapacitation. *Nature*, **400**, 449–52.

Prokopy, R. J., Diehl, S. R., and Cooley, S. S. (1988). Behavioral evidence for host races in *Rhagoletis pomonella* flies. *Oecologia*, **76**, 138–47.

Purcell, R. W. and Gould, S.J. (1993). *Finders, Keepers: Eight Collectors*. Pimlico, London.

Quammen, D. (1996). *The Song of the Dodo*. Pimlico, London.

Raijmann, L. E. L. (1996). In search for speciation: genetical differentiation and host race formation in *Yponomeuta padellus* (Lepidoptera: Yponomeutidae). Unpublished Ph.D. thesis, University of Amsterdam.

Rau, P. and Rau, N. (1929). The sex attraction and rhythmic periodicity in the giant saturniid moths. *Transactions of the Academy of Sciences of St Louis*, **26**, 83–221.

Raymond, M., Callaghan, A., Fort, P., and Pasteur, N. (1991). Worldwide migration of amplified insecticide resistance genes in mosquitoes. *Nature*, **350**, 151–3.

Reinthal, P. N. and Meyer, A. (1997). Molecular phylogenetic tests of speciation models in African cichlid fishes. In *Molecular Evolution and Adaptive Radiations*, (ed. T. J. Givnish and K. J. Sytsma), pp. 375–90. Cambridge University Press, Cambridge.

Rice, W. R. (1996). Sexually antagonistic male adaptation triggered by experimental arrest of female evolution. *Nature*, **361**, 232–4.

Rice, W. R. (1998). Intergenomic conflict, interlocus antagonistic coevolution, and the evolution of reproductive isolation. In *Endless Forms: Species and Speciation*, (ed. D. J. Howard and S. Berlocher), pp. 261–70. Oxford University Press, Oxford.

Rice, W. R. and Hostert, E. E. (1993). Laboratory experiments on speciation: what have we learned in 40 years? *Evolution*, **47**, 1637–53.

Rice, W. R. and Salt, G. W. (1988). Speciation via disruptive selection on habitat preference: experimental evidence. *The American Naturalist*, **131**, 911–17.

Rice, W. R. and Salt, G. W. (1990). The evolution of reproductive isolation as a correlated character under sympatric conditions: experimental evidence. *Evolution*, **44**, 1140–52.

Ridley, M. (1994). *The Red Queen: Sex and the Evolution of Human Nature*. Penguin, London.

Ritchie, M. (1997). Song divergence in the Zimbabwe strain of *Drosophila melanogaster*. In *Abstract Book of the Population Genetics Group Annual Meeting, Nottingham.*

Ringo, J., Wood, D., Rockwell, R., and Dowse, H. (1985). An experiment testing two hypotheses of speciation. *The American Naturalist*, **126**, 642–61.

Rodriguez-Trelles, F., Weinberg, J. R., and Ayala, F. J. (1996). Presumptive rapid speciation after a founder event in a laboratory population of *Nereis*: allozyme electrophoretic evidence does not support the hypothesis. *Evolution* **50**, 457–61.

Rosenzweig, M. L. (1995). *Species Diversity in Space and Time.* Cambridge University Press, Cambridge.

Rosenzweig, M. L. (1997). Tempo and mode of speciation. *Science*, **277**, 1622–3.

Rosenzweig, M. L. and McCord, R. D. (1991). Incumbent replacement: evidence for long-term evolutionary progress. *Paleobiology*, **17**, 202–13.

Rosenzweig, M. L. and Taylor, J. A. (1980). Speciation and diversity in Ordovician invertebrates: filling niches quickly and carefully. *Oikos*, **35**, 236–43.

Rundle, H. D., Mooers, A.Ø., and Whitlock, M. C. (1998). Single founder-flush events and the evolution of reproductive isolation. *Evolution*, **52**, 1850–5.

Ryan, M. J., Cocroft, R. B., and Wilczynski, W. (1990). The role of environmental selection in intraspecific divergence of mate recognition signals in the cricket frog, *Acris crepitans*. *Evolution*, **44**, 1869–72.

Schilthuizen, M. (1995). A comparative study of hybrid zones in the polytypic land snail *Albinaria hippolyti* (Pulmonata: Clausiliidae). *Netherlands Journal of Zoology*, **45**, 261–90.

Schilthuizen, M. and Lombaerts, M. (1994). Life on the edge: a hybrid zone in *Albinaria hippolyti* from Crete. *Biological Journal of the Linnean Society*, **54**, 111–38.

Schliewen, U. K., Tautz, D., and Pääbo, S. (1994). Sympatric speciation suggested by monophyly of crater lake cichlids. *Nature*, **368**, 629–32.

Schluter, D. (1994). Experimental evidence that competition promotes divergence in adaptive radiation. *Science*, **266**, 798–801.

Schluter, D. (1996). Ecological speciation in postglacial fishes. *Philosophical Transactions of the Royal Society*, **B351**, 807–14.

Schluter, D. (1998). Ecological causes of speciation. In *Endless Forms: Species and Speciation*, (ed. D. J. Howard and S. H. Berlocher), pp. 114–29. Oxford University Press, Oxford.

Schluter, D. and Nagel, L. M. (1995). Parallel speciation by natural selection. *The American Naturalist*, **146**, 292–301.

Schneider, C. J., Smith, T. B., Larison, B., and Moritz, C. (1999). A test of alternative models of diversification in tropical rainforests: ecological gradients versus rainforest refugia. *Proceedings of the National Academy of Sciences USA*, **96**, 13869–73.

Seehausen, O. (1996). *Lake Victoria Rock Cichlids: Taxonomy, Ecology, and Distribution*. Verduijn Cichlids, Zevenhuizen, the Netherlands.

Seehausen, O. (1999). Speciation and species richness in African cichlids: effects of sexual selection by mate choice. Unpublished Ph.D. thesis, Leiden University.

Seehausen, O. and van Alphen, J. J. M. (1998). The effect of male coloration on female mate choice in closely related Lake Victoria cichlids (*Haplochromis nyererei* complex). *Behavioural Ecology and Sociobiology*, **42**, 1–8.

Seehausen, O., van Alphen, J. J. M., and Witte, F. (1997). Cichlid fish diversity threatened by eutrophication that curbs sexual selection. *Science*, **277**, 1808–11.

Seehausen, O., Lippitsch, E., Bouton, N., and Zwennes, H. (1998). Mbipi, the rock-dwelling cichlids of Lake Victoria: description of three new genera and fifteen new species (Teleostei). *Ichthyological Exploration of Freshwaters*, **9**, 129–228.

Sepkoski, J. J. (1999). Rates of speciation in the fossil record. In *Evolution of Biological Diversity*, (ed. A. E. Magurran and R. M. May), pp. 260–82. Oxford University Press, Oxford.

Simpson, G. G. (1944). *Tempo and Mode in Evolution*. Columbia University Press, New York.

Smith, D. C. (1988). Heritable divergence of *Rhagoletis pomonella* host races by seasonal asynchrony. *Nature*, **336**, 66–7.

Smith, T. B., Wayne, R. K., Girman, D. J., and Bruford, M. W. (1997). A role for ecotones in generating rainforest biodiversity. *Science*, **276**, 1855–7.

Smith, T. B., Wayne, R. K., Girman, D., and Bruford, M. W. (in press). Evaluating the divergence-with-gene-flow model in natural populations: the importance of ecotones in rainforest speciation. In *Historical and Ecological Determinants of Diversity in Tropical Rainforests*, (ed. E. Bermingham and C. Moritz). University of Chicago Press, Chicago.

Soltis, D. E. and Soltis, P. S. (1999). Polyploidy: recurrent formation and genome evolution. *Trends in Ecology and Evolution*, **14**, 348–52.

Spurrier, M.F., Boyce, M. S., and Manly, B. F. J. (1991). Effects of parasites on mate choice by captive sage grouse. In *Ecology, Behavior and Evolution of Bird–Parasite Interactions*, (ed. J. E. Loye and M. Zuk). Oxford University Press, Oxford.

Stanley, S. M. (1979). *Macroevolution: Pattern and Process*. Freeman, San Francisco.

Stanley, S. M. (1982). Macroevolution and the fossil record. *Evolution*, **36**, 460–73.

Stebbins, G. L. (1950). *Variation and Evolution in Plants*. Columbia University Press, New York.

Sterk, A. A., Hommels, C. H., Jenniskens, M. J. P. J., Neuteboom, J. H., den Nijs, J. C. M., Oosterveld, P., and Segal, S. (1987). *Paardebloemen: Planten zonder Vader*. Stichting Uitgeverij KNNV, Utrecht.

Stiassny, M. L. J., Schliewen, U. K., and Dominey, W. J. (1992). A new species flock of cichlid fishes from Lake Bermin, Cameroon with a description of eight new species of *Tilapia* (Labroidei: Cichlidae). *Ichthyological Exploration of Freshwaters*, **3**, 311–46.

Strong, D. R., Lawton, J. H., and Southwood, R. (1984). *Insects on Plants. Community Patterns and Mechanisms*. Blackwell Scientific Publications, Oxford.

Stümpke, H. (1957). *Bau und Leben der Rhinogradentia*. Gustave Fischer, Stuttgart.

Sulloway, F. J. (1979). Geographic isolation in Darwin's thinking: the vicissitudes of a crucial idea. *Studies in the History of Biology*, **3**, 23–65.

Szymura, J. M. (1993). Analysis of hybrid zones with *Bombina*. In *Hybrid*

Zones and the Evolutionary Process, (ed. R. G. Harrison), pp. 261–89. Oxford University Press, Oxford.

Szymura, J. M. and Barton, N. H. (1986). Genetic analysis of a hybrid zone between the fire-bellied toads *Bombina bombina* and *B. variegata*, near Cracow in Southern Poland. *Evolution* **40**, 1141–59.

Szymura, J. M. and Barton, N. H. (1991). The genetic structure of the hybrid zone between the fire-bellied toads *Bombina bombina* and *B. variegata*: comparisons between transects and between loci. *Evolution* **45**, 237–61.

Tatarenkov, A. and Johannesson, K. (1998). Evidence of a reproductive barrier between two forms of the marine periwinkle *Littorina fabalis* (Gastropoda). *Biological Journal of the Linnean Society*, **63**, 349–65.

Tattersall, I. (2000), Paleoanthropology: the last half century. *Evolutionary Anthropology*, **9**, 2–16.

Tauber, C. A. and Tauber, M. J. (1989). Sympatric speciation in insects: perception and perspective. In *Speciation and its Consequences*, (ed. D. Otte and J. A. Endler), pp. 307–344. Sinauer, Sunderland, MA.

Taylor, E. B., Foote , C. J., and Wood, C. C. (1996). Molecular genetic evidence for parallel life-history evolution within a Pacific salmon (sockeye salmon and kokanee, *Oncorhynchus nerka*). *Evolution*, **50**, 401–16.

Templeton, A. R. (1979). The unit of selection in *Drosophila mercatorum*. II. Genetic revolution and the origin of coadapted genomes in parthenogenetic strains. *Genetics*, **92**,1265–82.

Templeton, A. R. (1989). Founder effects and the evolution of reproductive isolation. In *Genetics, Speciation and the Founder Principle*, (ed. L. V. Giddings, K. Y. Kaneshiro, and W. W. Anderson), pp. 329–344. Oxford University Press, Oxford.

Templeton, A. R. (1996). Experimental evidence for the genetic-transilience model of speciation. *Evolution*, **50**, 909–15.

Thomas, S. and Singh, R. S. (1992). A comprehensive study of genetic variation in natural populations of *Drosophila melanogaster*. VII. Varying rates of genetic divergence as revealed by two-dimensional electrophoresis. *Molecular Biology and Evolution*, **9**, 507–25.

Ting, C.-T., Tsaur, S.-C., Wu, M.-L., and Wu, C.-I. (1998). A rapidly evolving homeobox at the site of a hybrid sterility gene. *Science*, **282**, 1501–4.

Turelli, M. and Hoffmann, A. A. (1991). Rapid spread of an inherited incompatibility factor in Californian *Drosophila*. *Nature*, **353**, 440–2.

Turner, G. E. and Burrows, M. T. (1995). A model of sympatric speciation by sexual selection. *Proceedings of the Royal Society of London*, **B260**, 287–92.

Tweedie, T. W. F. (1961). On certain Mollusca of the Malayan limestone hills. *Bulletin of the Raffles Museum*, **26**, 49–65.

Van Valen, L. (1973). A new evolutionary law. *Evolutionary Theory*, **1**, 1–30.

Vincek, V., O'Huigin, C., Satta, Y., Takahata, N., Boag, P. T., Grant, P. R., Grant, B. R., and Klein, J. (1997). How large was the founding population of Darwin's finches? *Proceedings of the Royal Society of London*, **B264**, 111–18.

Vrba, E. S., Denton, G. H., Partridge, T. C., and Burckle, L. H. (ed.) (1995). *Paleoclimate and Human Evolution, with Emphasis on Human Origins*. Yale University Press, New Haven, CT.

Vries, H. de (1901). *Die Mutationstheorie*; Vol. 1. Von Veit Verlag, Leipzig.

Wagner, M. (1868). *Die Darwin'sche Theorie und das Migrationsgesetz der Organismen*. Duncker & Humblot, Leipzig.

Wagner, M. (1889). *Die Entstehung der Arten durch räumliche Sonderung*. Schwalbe, Basel.

Wallace, A. R. (1869). *The Malay Archipelago. The Land of the Orang-Utan and the Bird of Paradise: A Narrative of Travel with Studies of Man and Nature*. Macmillan, London.

Walsh, B. J. (1864). On phytophagous varieties and phytophagous species. *Proceedings of the Entomological Society of Philadelphia*, **3**, 403–30.

Walsh, B. J. (1867). The apple-worm and the apple maggot. *Journal of Horticulture*, **2**, 338–43.

Waring, G. L., Abrahamson, W. G., and Howard, D. J. (1990). Genetic differentiation among host-associated populations of the gallmaker *Eurosta solidaginis* (Diptera: Tephritidae). *Evolution*, **44**, 1648–55.

Weinberg, J. R., Starczak, V. R., and Jörg, D. (1992). Evidence for rapid speciation following a founder event in the laboratory. *Evolution*, **46**, 1214–20.

Wilkens, H. (1993). Neutral mutation and evolutionary progress. *Zeitschrift für Zoologische Systematik und Evolutionsforschung*, **31**, 98–109.

Williamson, P. G. (1981). Paleontological documentation of speciation in Cenozoic molluscs from Turkana Basin. *Nature*, **293**, 437–43.

Wilson, E. O. (1992). *The Diversity of Life*. Penguin Books, London.

Wirtz, P. (1978). The behaviour of the Mediterranean *Tripterygion* species (Pisces, Blennioidei). *Zeitschrift für Tierpsychologie*, **48**, 142–74.

Wood, T. K. (1993). Speciation of the *Enchenopa binotata* complex (Insects: Homoptera: Membracidae). In *Evolutionary Patterns and Processes*, pp. 299–318. The Linnean Society, London.

Wood, T. K. and Guttman, S. I. (1983). *Enchenopa binotata* complex: sympatric speciation? *Science*, **220**, 310–12.

Wood, T. K. and Keese, M. C. (1990). Host-plant induced assortative mating in *Enchenopa* treehoppers. *Evolution*, **44**, 619–28.

Wright, S. (1931). Evolution in Mendelian populations. *Genetics*, **16**, 97–159.

Wu, C.-I., Johnson, N. A., and Palopoli, M. F. (1996). Haldane's Rule and its legacy: why are there so many sterile males? *Trends in Ecology and Evolution*, **11**, 281–4.

Glossary

adaptation The evolution of characteristics that make an organism fit for the environment it lives in. Sometimes the term also refers to such a characteristic itself, rather than to the process.

allele A version of a gene.

allopatric speciation The evolution of a new species in an isolated area.

allopatry The presence of two populations or species in two geographically separate areas. When new species evolve in allopatry, they undergo allopatric speciation.

anatomy The study of the inner organs of an animal or plant.

banana flies Small flies of the family Drosophilidae, often used in biological research. The more common name for these flies is 'fruit flies', but this name is more properly applied to the 'real' fruit flies, Tephritidae, to which the apple maggot fly belongs.

benthic Living near the bottom of water bodies; for contrast, *see* limnetic.

biogeography The science of charting and explaining the global distribution of animals and plants.

biological species concept (BSC) Ernst Mayr's definition, that species are populations that are sexually isolated from other populations. The most commonly used, but not always very helpful, definition of species.

biotope A particular type of environment; for example, temporary puddles of rain water and large permanent ponds are different biotopes.

bottleneck A period when a population is reduced to a very small number. This can result in drastic changes in the genetic composition of the population, due to sheer chance. The founder effect, for example, is caused by a bottleneck (see below).

chloroplasts The green blobs in a plant cell in which photosynthesis takes place; because chloroplasts evolved from bacteria, they have their own DNA.

chromosome A stretch of DNA on which many genes lie.

cichlids A family of freshwater fish (the Cichlidae) that contains thousands of different species. They figure quite prominently in the literature about speciation.

clone An individual that is genetically identical to another individual; identical twins are clones, for example.

computer simulation Using a computer to replay a process from the real world. A common procedure to see if a new theoretical idea is viable, without having to do any tedious experiments.

diapause A resting stage in the development of, for example, insects.

diploid With a double set of chromosomes, one set from each parent; the normal situation.

DNA Deoxyribonucleic acid—the stuff that carries the genetic code.

DNA fingerprinting A technique for quickly estimating genetic relatedness.

endemic Occurring only in one spot. The dodo, for example, was endemic to the island of Mauritius: it evolved, lived, and expired there.

enzyme A protein that carries out a chemical reaction.

ecology The branch of biology that studies the relationships between an organism and its environment. Sometimes also used as a name for the environmental movement, but this is incorrect—many ecologists are environmentalists but not all environmentalists are ecologists.

ecotone An area where one type of environment changes into another. The edge of a rain forest, for example, is an ecotone.

entomology The study of insects.

female choice The preference of females for certain types of males. Once considered a ridiculous idea, it is now universally accepted.

Fisherian sexual selection The evolution of exaggerated male ornaments because of female choice.

founder effect The effect that a small founder colony can give rise a population that is genetically different from the parent population, due to chance.

founder–flush speciation Hampton Carson's theory of the evolution of new species due to the founder effect followed by a rapid population growth (the flush).

fruit flies A family of flies that usually feed and live on fruit, the Tephritidae. Not to be confused with banana flies, which are also often called by this name (see above).

gene A region on a chromosome that carries the genetic code for a protein and, ultimately, for a characteristic of the organism.

gene flow The result of mating and migration, gene flow spreads genes around a population, among different populations, or even among different species.

gene pool All the genes and their alleles that are in a population. Gene flow mixes the gene pool.

genetic drift Changes in the genetic composition of a population due to chance. It is only important in small populations.

genetic isolation The same as reproductive isolation: absence of gene flow between two populations.

genetics The branch of biology that studies anything that has to do with genes (and that is a lot).

genitalia The 'primary' sexual organs of animals; usually this means the penis and the vagina, or their equivalents.

genome The entire collection of genes of a species and the way they are organized on the chromosomes.

genus A group of closely related species, recognized in the first part of a species name; in *Drosophila melanogaster* (the laboratory banana fly), for example, *Drosophila* is the genus, which includes more than a thousand other species. It is a bit like people's names: the surname is the genus, showing who is related to whom, while the first name is the species name; first name and surname together indicate a unique person, just like genus and species name together refer to a unique species.

geographical speciation The same as allopatric speciation.

geographical isolation When animals of plants cannot meet and mate due to an insurmountable geographical barrier, such as a mountain range, a sea, a glacier, or a desert.

glaciation The expansion of glaciers during the Ice Ages.

good-genes The theory that female choice is not just aesthetic, but based on characteristics that convey genetic quality, such as a good immune system.

herbivorous Plant-eating.

herpetology The study of reptiles.

host race A form of a species that is specialized on a particular food plant. Many host races can just as easily be viewed as different species.

host shift This term is mostly used in reference to plant-feeding or parasitic animals that change their host (in evolution). When the apple-feeding banana fly evolved from a hawthorn-feeding one, this involved a host-shift.

hybrid The genetically mixed offspring of a mating between two different species, subspecies, or host races. Hybrids can also be the offspring of other hybrids. Often (but by no means always) hybrids are inviable, sterile, or in other ways worse off.

hybrid zone An area where two different species, subspecies, or host races meet, mate, and interbreed.

invertebrate An animal with no backbone. Insects, snails, worms, and most other creepy-crawlies are invertebrates; mammals, birds, amphibians, reptiles and fish, on the other hand, are vertebrates.

lek Derived from the Swedish word for 'play', a lek is the ground where many males display for passing females. Leks are mostly used by birds and insects.

limnetic Living near the surface of water bodies (as opposed to benthic).

malacology The study of molluscs (that is, shells, snails, squid, and octopuses).

macroevolution Big-time evolution. The origin of entirely new classes of animals and plants; for example, the evolution of birds or of flowering plants. Speciation is sometimes considered macroevolution, but usually not.

microbiology The branch of biology that deals with microbes (usually including all organisms that are too small to see—bacteria, viruses, yeasts).

microevolution Peanuts evolution. Evolutionary change within a species; for example, the evolution of black coloration in moths in industrial areas. Speciation is usually considered microevolution, but sometimes not.

mitochondria Small blobs inside the cells of animals and plants that produce the cell's energy. Because mitochondria (singular, mitochondrion) evolved from bacteria, they still carry their own DNA.

Modern Synthesis The biological revolution that took place in the 1930s and 1940s. Combining taxonomy, palaeontology, genetics, and Darwinism, it paved the way for a more integrative biology. Also known as the New Synthesis.

monogamy Animals are monogamous if they mate for life. If not, they are polygamous. Obviously, the distinction is not so strict: there are polygamous sorts of monogamy and monogamous sorts of polygamy. People are usually considered to sit on the monogamous end of the animal kingdom, but, if this were true, the Seventh Commandment would have been superfluous.

monophagy Specialization on a single species of food plant.

morphology The study of the shape of animals and plants; for example, morphologically, male and female stag beetles are very different, but, genetically, they are very similar.

mutant An individual that carries a mutation.

mutation The process or the end result of a genetic accident; this could range from a change in a single letter in the DNA code of a gene, or the duplication of the entire genome. Mutations can be anything from harmful to beneficial; if beneficial, natural selection will make them spread.

natural selection Charles Darwin and Alfred Russel Wallace discovered this. It is the process by which evolution works. Those individuals that are genetically the best suited to their environment will survive, so they will produce the most offspring. Hence, in the next generation, their genes will be present at a higher frequency.

niche A species' 'occupation'; its place in the ecological network.

ornithology The study of birds.

oviposition Egg-laying.

ovipositor The organ a female uses for depositing her eggs (usually present when eggs are injected into a substrate such as fruit, soil, or wood).

ovule Basically this is the egg cell in a seed-bearing plant, which is fertilized by pollen.

palaeontology The study of fossils.

parthenogenesis Virgin birth; producing offspring without fertilization.

peripheral isolates Small populations at the edge of a species' range. The term is also used for Ernst Mayr's theory of speciation, which relied particularly on these small colonies.

pleiotropy When a single gene has more than a single effect; for example, a change in a 'clock gene' can offset many biological rhythms.

Pleistocene epoch The period between about 1.9 million and 10 000 years ago, which was characterized by repeated cycles of cold climate (the so-called Ice Ages, or glacials) interspersed by warmer interglacials.

pollen The minute grains produced by plants' anthers that contain the male sex cells and fertilize ovules.

pollination The process of fertilization in plants. Because plants cannot move, they often need insects as genital proxies.

polygamy Having more than one sexual partner. *See* monogamy.

polyploid An individual with more than two sets of chromosomes. This can happen when one of the sex cells from which it sprang 'forgot' to halve its genome.

population All the individuals of a species of plants or animals that live in a certain place.

population genetics The science that studies changes in frequencies of alleles in populations. Genetic drift, natural selection, and gene flow all fall within the province of population genetics. Many people (but not palaeontologists) consider population genetics as synonymous with evolutionary biology.

postglacial lake A lake that formed after the last ice age, which makes it less than 10 000 years old.

post-mating isolation When two individuals cannot have any offspring because something goes awry after mating. This could be the inability of sperm to reach the egg, the death of the eggs or embryos, or some other mishap.

pre-mating isolation When two individuals cannot have any offspring because something goes awry before mating takes place. For example, the two potential partners cannot find each other because they are active at different times, because they use different sexual signals, or because mating itself is impeded (snails that are one another's mirror image, genitalia that do not match, and so on).

protein A large molecule made up of hundreds of amino acids, the exact make-up of which is encoded in a gene. Most functions in living organisms are carried out by proteins.

punctuated equilibria The view, based on the fossil record, that evolution proceeds in fits and starts—long periods of stability are suddenly punctuated by grand changes.

recombination The genetic jumble that takes place when sex cells are formed. It involves the mixing of genes from both parents.

reproductive isolation The inability of two individuals (or species) to interbreed, due to pre-mating isolation, post-mating isolation, or both (see above).

runaway sexual selection Fisherian sexual selection (see above).

sensillum A hair or peg, usually on insects' antennae, that is sensitive to smells (plural, sensillae).

selection Usually natural selection is meant.

selection pressure The severity of natural selection.

sex cells Sperm and pollen in male animals and plants, respectively; and eggs and ovules in female animals and plants, respectively. These are the cells that, after recombination, carry only one set of chromosomes and are supposed to fuse with a sex cell from another individual to produce a new individual.

sexual selection Darwin's second great discovery. The process of evolution where not the environment but the opposite gender does the selecting. Males especially are prone to show the effects of sexual selection.

speciation The evolution of a new species.

species No easy answer possible—see Chapter 1.

speleology Cave exploration.

sperm The sex cells of male animals.

subspecies The same as a species, but still interbreeding with another subspecies.

sympatric speciation The evolution of a new species without geographical isolation.

sympatry The presence of two species within the same geographical area.

taxonomy The branch of biology whose business it is to find, describe, and classify species.

tetraploid Having four sets of chromosomes, rather than the usual two. Tetraploids usually form when two (diploid) sex cells accidentally neglect to halve their double set of chromosomes and fuse.

vertebrate An animal with a backbone. *See* invertebrate (above).

Index

Numbers in italics indicate illustrations